Cultural Landscapes of Israel

Aviad Sar Shalom • Yuval Peled • Rachel Singer •
Irit Amit-Cohen • Rafi Rich • Avraham (Avi)
Sasson • Elissa Rosenberg

Cultural Landscapes of Israel

Springer

Aviad Sar Shalom
Tel Aviv, Israel

Rachel Singer
Bezalel Academy of Arts and Design
Jerusalem, Israel

Rafi Rich
Architect
Petach Tikva, Israel

Elissa Rosenberg
Bezalel Academy of Arts and Design
Jerusalem, Israel

Yuval Peled
Division of Planning and Development
Israel Nature and Parks Authority
Jerusalem, Israel

Irit Amit-Cohen
Department of Geography and Environment
Bar Ilan University
Ramat Gan, Israel

Avraham (Avi) Sasson
Ashkelon Academic College
Kvutzat Yavne, Israel

The original Hebrew publication was supported by the Israel Nature and Parks Authority, the Beracha Foundation and the Israel National Commission for UNESCO.

ISBN 978-3-031-33687-4 ISBN 978-3-031-33685-0 (eBook)
https://doi.org/10.1007/978-3-031-33685-0

Cover illustration: A view of Mount Tabor, the setting of Biblical events and the site of the Transfiguration of Jesus, with the village of Daburiyya in the foreground. Mount Tabor is an associative landscape on the World Heritage Tentative List as a component part of *The Galilee Journeys of Jesus and the Apostles*, anchored by the Greek Orthodox and the Franciscan Churches of the Transfiguration. Photo: Itamar Grinberg.

This Springer imprint is published by the registered company Springer Nature Switzerland AG
The registered company address is: Gewerbestrasse 11, 6330 Cham, Switzerland

Foreword

In January 2020, at the request of the Israel Nature and Parks Authority, I engaged in discussions with Israeli experts about the important cultural landscapes of Israel. Our tours of the Sea of Galilee, Bet Netofa Valley, Jerusalem terraces, Refaim and Ela valleys, and Ramat Hanadiv revealed a wealth of places, times, and stories that were a delight to explore. We studied the landscapes of David and Goliath and the Battle of Hattin, where geography illuminates the historical events.

Interestingly, given the relatively small size of Israel, there are four different climatic and ecological biomes embedding variations in landscapes and waterscapes into cultural landscapes. These significant cultural land and seascapes bear ancient traces of the early movements of people below and above ground, hold evidence of specific historical times and ways of life, and reveal imprints of recent generations. They reflect religious and philosophical beliefs and events or bear witness to agricultural practices and social movements, serving as vessels for meaningful stories and places of traditions. These diverse cultural landscapes arose across a broad period and were shaped by humanity and nature to differing degrees, suiting place-based purposes.

When cultural landscapes were approved for listing as World Heritage assets in 1992, they were identified as three basic types: 1. defined and designed at a point in time; 2. evolved over the years and currently in relict form or continuing to evolve; and 3. associated with beliefs, arts, literature or inspiration. As of 2023, some 121 cultural landscapes have been inscribed on the World Heritage list, with the majority categorized as evolved continuing landscapes. Israel's sole World Heritage cultural landscape inscription is the Incense Route—Desert Cities of the Negev. Most of Israel's cultural landscapes are in the evolved relict category, arising from the eons of evolution across these lands on the east coast of the Mediterranean Sea. These land and sea cultural assets are of universal, national, and local significance.

Today, Israel's cultural landscapes are little known globally. With its publication in English, this book, *Cultural Landscapes of Israel*, brings the most important of these combined works of humanity and nature to a broader audience. The publication of this detailed exploration of cultural landscapes is the cause for celebration as it comes forward on a rising tide of activism and recognition of their value to the people of Israel and their visitors. Globally protected landscapes accrue diverse, sustainable benefits for culture, economy, society, and environment. Israel's unique cultural landscapes and seascapes must be more broadly appreciated and valued as the multiple pressures of economic growth, social evolution, and environmental challenges converge to make these places even more critical as anchors for contemporary society and the generations to come. I commend this excellent work on the cultural landscapes of Israel to worldwide readers with the secure knowledge that each of you will learn much and enjoy the journey.

Patricia M. O'Donnell, PLA, FASLA, AICP, F. US/ICOMOS
President, ICOMOS IFLA International Scientific
Committee on Cultural Landscapes

Preface

Over 10 years have passed since *Cultural Landscapes of Israel* was published in Hebrew. The goal of that publication was to define the emerging concept of cultural landscapes and promote their protection in Israel, a land with a richly layered, diverse heritage. Rapid population growth in Israel has accelerated the pace of development, putting pressure on natural and cultural resources alike. The ongoing threats to both landscape resources and cultural assets have highlighted the interdependence of the natural environment and cultural heritage, stressing the importance of bringing these two realms into a single conceptual framework. Thus, the concept of the cultural landscape, the encounter between human and natural environments, was developed as a planning tool in Israel. The 2010 edition of *Cultural Landscapes of Israel* sought to introduce these concepts to the Israeli public and adapt them to the local planning context. It proposed a preliminary series of designations following the guidelines adopted by UNESCO's World Heritage Committee (1992) and elaborated in subsequent documents.

This newly translated volume aims to make this work accessible to a broad international audience. We believe that Israel, a land with a deep and varied history and a remarkably diverse range of physiographic regions and landscape types, has compelling lessons to offer the study of cultural landscapes. It is a land where biblical narratives from both the Old and New Testaments still resonate through its landscapes, making the project of cultural landscape protection a unique yet urgent challenge and an important complement to its rich archaeological and natural heritage.

Conflict is perhaps the most salient feature of Israeli cultural landscapes, where religious or nationalistic narratives often shape identity. Israel is characterized by its cultural and religious diversity, multiple histories, and competing narratives. At the same time, its geography accounts for its ancient role as a regional crossroads, a bridge across cultures. Because of this diversity, it was vital to establish a common, unifying methodology when establishing an approach to defining Israel's cultural landscapes. It was important to stress that the local natural, social, and cultural attributes may reflect a multitude of experiences. For these reasons, the World Heritage criteria for cultural landscapes were adopted and adapted to the unique characteristics of Israel. This allowed for the recognition of a range of associations and meanings embedded in the landscape.

Yet, the cultural landscape is an evolving concept with many meanings. Since the initial publication of the Hebrew edition in 2010, the field has continued to develop, bringing new issues to the fore. We have updated the key concepts in the current volume to reflect the changes that have occurred in the international discourse. Some of these include the growing environmental awareness in light of climate change; the role of cultural landscapes in safeguarding food sources and sustainable agricultural practices related to food security; the continued rise of globalized economies; changing relationships between cultural landscapes and open space for leisure and recreation; and approaching cultural landscapes as productive sources of livelihoods, an idea that is central to their sustainability and management.

Planning methodologies have also been evolving. These have been shaped by the expanded use of digital technologies and perspectives that stress the social role of cultural landscape preservation. The Sustainable Development Goals established by the United Nations in 2015 have stressed the importance of advancing dialogue and creating new stakeholder partnerships to strengthen community involvement. This community focus also serves the larger purpose of contributing to cultural diversity.

Moreover, since 2010, Israel's awareness of the significance of cultural landscapes has grown. The publication of the Hebrew edition of the book succeeded in forging a broad consensus regarding the definition of cultural landscapes as established in the international literature. These have become absorbed into the planning discourse and practice at national, district, and local planning levels in Israel and adopted by a wide array of public agencies and local authorities. And yet, more work must be done to advance the protection of cultural landscapes by integrating them into the planning processes using a robust regulatory system.

The first chapter provides a brief theoretical overview of cultural landscapes, situating the extensive work conducted on cultural landscapes in Israel within a larger global and scholarly context. It presents several interpretations and approaches, exploring the significance of the idea of contested landscapes for heritage management in Israel.

Chapter 2 describes the international frameworks for protecting cultural landscapes. It reviews the inscription process and other mechanisms used to safeguard landscapes worldwide, concluding those with directives recommended for Israel.

The Israeli context is presented in Chap. 3 from cultural and natural perspectives. The land's unique physical attributes are outlined with a brief survey of its geomorphology and a discussion of its diverse biomes. Similarly, Israel's cultural and historical complexity is discussed by examining the key narratives embedded in Israel's landscapes and explaining the planning principles used for mapping its cultural landscapes.

Chapter 4 reviews the relevant physical planning mechanisms in Israel with an emphasis on the National Outline Plans, Thematic Plans, and District Plans to understand their impact on landscape protection and the various stakeholders involved. The methodology for defining cultural landscape polygons is presented with the cultural anchors that situate key points in the site and their buffer zones that afford an additional layer of protection.

A collection of 50 cultural landscape case studies is presented in Chap. 5. The sites are classified according to their level of significance and divided into cultural landscape categories. The list begins with the cultural landscapes inscribed on the World Heritage List, followed by those inscribed on the World Heritage Tentative List. It then lists cultural landscapes with a national level of significance. These are subdivided into the three categories established by the World Heritage Committee: designed landscapes, organically evolved landscapes (both relict and continuing landscapes), and associative landscapes.

Chapter 6 focuses on cultural landscape management mechanisms as illustrated by three innovative case studies that stress the role of public participation, a World Heritage site, a Biosphere Reserve, and a National Park.

This volume is but a first step in introducing a broad audience to the rich resources and challenges of Israel's cultural landscapes.

Ramat Gan, Israel Irit Amit-Cohen

Irit Amit-Cohen Professor Emerita, Department of Geography and Environment, Bar Ilan University, coordinated for many years the Program on Conservation Planning and Development of Cultural Landscape and Cultural Heritage. She is a member of ICOMOS Israel and the Board of Directors of the Council for Conservation of Heritage Sites in Israel.

Acknowledgements

The recognition of cultural landscapes in Israel has resulted from many years of public activism, professional development, and academic research by passionate persons committed to environmental protection. We have received support and comments from many friends and colleagues. We wish to thank the late Sara Arenson, Doron Bar, Dorit Eilon, Moshe Hartal, Motti Kaplan, Moshe Lerner, Menachem Marcus, Osnat Rosen-Kramer, Rehav Rubin, Yaakov Shkolnik, Yuval Shuali, and Yossi Stefanski for materials, suggestions, and support throughout the extensive research process that led to the establishment of the current cultural landscapes collection.

We further wish to acknowledge Professor Nurit Lissovsky for her contribution to the section on Sachne (Gan Hashlosha), Ben Gurion's Tomb, and Tombs of the Righteous (Tzadikim) in Chap. 5, Dr. Rafael Y. Lewis for the Horns of Hattin entry and Rafi Rich for his contribution on the statutory planning framework in Israel. We also acknowledge Avraham (Avi) Sasson's contribution to the educational aspects of cultural landscapes in the earlier Hebrew edition of this book.

Advisory Committee: Yuval Peled (Chair), Ran Aharonson (Department of Geography, Hebrew University), Tal Alon-Mozes (Faculty of Architecture and Town Planning, Technion-Israel Institute of Technology) Sefi Beeri, Daniel Bar-Elli Bitton, Yehuda Dagan (Conservation Department, Israel Antiquities Authority), Zeev Margalit (Israel Nature and Parks Authority) Rehav Rubin (Department of Geography, Hebrew University), Anat Sadeh (Anat Sadeh, Landscape Architecture), Lea Shneor (Lea Shneor Architect—Architecture & Town Planning), Tami Trop (The Natural Resources and Environmental Research Center, University of Haifa), Mike Turner (UNESCO Chair in Urban Design and Conservation Studies, Bezalel Academy of Arts and Design), Tzvika Tzuk (Israel Nature and Parks Authority), Menachem Zalutzki (Ministry of Environment) and Yehuda Ziv.

Mapping and computing: The Center for Geographic-Ecologic Information, The Department of ICT and Information Systems, GIS unit, Israel Nature and Parks Authority.

Diagrams: Gila Tzionit, The Center for Geographic-Ecologic Information, The Department of ICT and Information Systems, GIS unit, Israel Nature and Parks Authority.

Translation and editing: Rachel Singer, Miriam Fierberg, Elissa Rosenberg.

Photography: Itamar Grinberg, Israel Nature and Parks Authority Archive.

Contents

Abbreviations

IAA	Israel Antiquities Authority
ICOMOS	International Council on Monuments and Sites
INPA	Israel Nature and Parks Authority
IUCN	International Union for Conservation of Nature
JCA	Jewish Colonization Association
JNF	Jewish National Fund
OG	Operational Guidelines for the Implementation of the World Heritage Convention
PICA	Palestine Jewish Colonization Association
SPNI	Society for the Protection of Nature in Israel
WH	World Heritage

List of Figures

1 Cultural Landscapes—Humans and Nature

The primary focus of this book is to introduce an inventory of proposed cultural landscapes in Israel that have been identified, researched, and mapped by the Israel Nature and Parks Authority. The first step of identifying cultural landscapes in Israel was based on adapting the definitions developed in the international conventions to the Israeli context. A second goal was to stress the local attributes of natural, social, and cultural values and to reflect a broad range of experiences.

Before we examine the definitions and criteria developed by the World Heritage Convention and understand how they have been interpreted and implemented within the local Israeli planning context, we begin by tracing the conceptual roots of the idea of cultural landscapes and its evolution. This chapter outlines a set of conceptual frameworks that have shaped contemporary approaches to cultural landscape conservation that are especially relevant to Israel.

1.1 Conceptual Frameworks

One of the earliest references to cultural landscapes dates to 1908 when German geographer Otto Schlüter distinguished between an "original landscape" (*Urlandschaft*) present before human intervention and a "cultural landscape" (*Kulturlandschaft*) that was a product of human culture (Elkins, 1989). The American geographer Carl Sauer is often credited with developing the meaning of the term and promoting its use in the 1920s and 1930s. In his book *The Morphology of Landscape* (1925), he defined a cultural landscape as a natural landscape modified by a cultural group. He wrote: "the cultural landscape is fashioned from a natural landscape by a culture group. Culture is the agent, the natural area is the medium, the cultural landscape is the result" (1925, p. 46). To study the landscape, one must analyze both its physical and cultural foundations.

J. B. Jackson further expanded the idea of the cultural landscape. A public intellectual and a prolific essayist, Jackson was a great observer of the overlooked, everyday spaces of the American landscape. His influential essays brought these spaces into focus—fences, backyards, sidewalks, parking lots, motels, mobile homes—and decoded these ubiquitous sites as sources of myth and symbolic meaning. He did not view landscape as a work of art, as a natural space, or as "scenery," but as "a composition of man-made or man-modified space to serve as infrastructure or background for our collective existence" (Jackson, 1986, p. 8). Jackson's cultural emphasis challenged purely aesthetic approaches to landscape. "No landscape, vernacular or otherwise, can be comprehended unless we perceive it as an organization of space; unless we ask ourselves, who owns or uses those spaces, how there were created and how they change" (Jackson, 1986, p. 147). However banal or quotidian, all landscapes were worthy objects of study: "There is really no such thing as a dull landscape or farm or town. A rich and beautiful book is always open before us. We have but to learn to read it" (cited in Groth et al., 1997, p. 3). Jackson opened up new subjects for study based on the belief that everyday experience is essential to forming human meaning and community identity.

While Sauer's understanding of the landscape as the product of human activity has remained a foundation of cultural landscape studies, his approach was later critiqued by the new cultural geographers who challenged his notion of culture as overly mechanistic. Geographer James Duncan (1990) and others questioned Sauer's conception of culture as external to the individual, thereby denying human agency. This critique emphasized the need to closely analyze the

A. Sar Shalom et al., *Cultural Landscapes of Israel*,
https://doi.org/10.1007/978-3-031-33685-0_1

workings of cultural processes and the active role played by individuals and groups, thereby stressing the importance of landscape change and the underlying power relationships implicit in that change (Schein, 1997). The approach to cultural landscapes in the 1980s and 1990s engaged with political and social issues, proposing a more nuanced and critical approach. Instead of viewing culture as a monolithic force, critics sought to address the broader social context in which cultures are expressed (Jackson, 1986), including issues such as conflict, social difference, inequality, and exclusion. Landscape could not be studied as a static "final product" of human activity. Instead, it was proposed as a conceptual framing that actively shaped social life and the performance of identity (Duncan & Duncan, 2001; Schein, 1997; Wylie, 2007). Alongside the descriptive, material focus on the artifact that grew out of Sauer's empirical approach, cultural landscapes began to address non-material, symbolic dimensions (Jackson, 1989 in Wylie, 2007). The concept further developed within distinct, sometimes overlapping frameworks that viewed the landscape as a "text" or a cultural resource, and variously emphasized their contested meanings, their relationship to landscape change and, more specifically, climate change, and the significance of community engagement. This has been reflected in the deliberations of the World Heritage Committee in conditioning nominations with the approval of management systems and a greater awareness of cultural diversities and heritage dissonance.

1.1.1 Landscape as Text

Duncan and others' interest in the symbolic qualities of landscape gave rise to the idea of landscape as a text that had to be "read" and interpreted (Cosgrove & Jackson, 1987; Duncan, 1990). Landscape, in this view, is not just a historical record; it plays an ideological role by supporting certain valued or unquestioned assumptions (Cosgrove & Jackson, 1987; Duncan, 1990). Reading the landscape is a way of interpreting it and decoding its underlying assumptions and meanings. Thus, the act of "reading" is part of a critical agenda meant to expose "the mechanisms through which dominant ideas and beliefs are reproduced via landscapes" (Wylie, 2007, p. 72). Texts contain multiple, often contradictory meanings (Barnes & Duncan, 1992). Landscape-as-text stresses the representational character of the landscape.

The landscape's representational role has also been studied from other perspectives based on the following key principles. First, the history of the landscape is based on its documentation, its physical evolution (its geology and geomorphology), and its associated historical events (Everson & Williamson, 1998; Hooke, 1998; Muir, 1999). The second principle states that heritage, memory, myth, and symbols related to landscape are not fixed but change over time and vary across communities (Hobsbawm & Terence, 1983). The third principle relates to the processes of selection, commemoration, representation and design, and the agendas they serve (Azaryahu, 1995; Daniels, 1993; Jackson, 1994; Meinig, 1979; Shamir, 1996), suggesting that the representational role of the landscape allows it to be used to serve diverse purposes. It can be used to strengthen a community's identification with its surrounding environment on both the local and national levels (Amit-Cohen, 2005; Ashworth, 1994) or to convey hegemonic values and interests (Duncan, 1990; Osborne, 1998). Finally, the representational role of landscape suggests that it should be analyzed not only based on its visible characteristics but also on its latent and experiential qualities.

1.1.2 Landscape as Cultural Resource

In addition to the representational role of landscape, recent research has increasingly stressed the economic significance of cultural landscapes, regarding them as consumer products that behave according to market forces. As a result, landscapes may be studied using parameters such as demand, adaptability to current needs and functions, the cost of adaptation, the added value that the landscape or the cultural resource provides to the selected function, and the constraints to development and their solution. Other aspects include its role in providing livelihoods, well-being, and recreation. While these issues have been studied typically in relation to tourist attractions, commercial, leisure, and recreational infrastructure (Ashworth & Dehan, 1986; Ashworth & Larkham, 1994), an expanding group of landscape types has been studied from an economic perspective, including those used for religious or cultural gatherings or educational activities (Amit-Cohen, 2005; Amit-Cohen & Amram, 2002; Featherstone, 2007; Hewison, 1987). The landscape's aesthetic characteristics and inspirational values are often used as a marketing tool. The commodification of traditional sites solely for tourist purposes can be problematic, complicating how these sites are narrated and managed and which voices are privileged (AlSayyad, 2001).

1.1.3 Resilience and Landscape Change

Sustainability and resilience are critical frameworks for identifying and managing cultural landscapes. Although landscapes are constantly evolving and adapting to change, the "speed, scale and magnitude of landscape and ecosystem change in the past 50–60 years have been unprecedented" (Plieninger & Bieling, 2012, p. 8). This global situation is even more acute in Israel, which has been facing accelerated urbanization and the loss of rural land for many decades. The intensification of agricultural practices also accounts for environmental degradation worldwide, including declining biodiversity, increased habitat fragmentation, degraded water quality, reduced forest cover, and increased greenhouse gases (Millennium Ecosystem Assessment, 2005). Some scholars have argued for using a resilience framework for understanding cultural landscapes and the drivers of landscape change to manage them more strategically (Aktürk & Dastgerdi, 2021; Plieninger & Bieling, 2012). The ecosystem service approach provides another framework for analyzing and managing cultural landscapes. Based on the Millenium Ecosystem Assessment (MEA) in 2005, this approach identifies and promotes the benefits that ecosystems contribute to human well-being through four basic service types: provisioning services (such as food and clean water), regulating services (such as disease prevention or flood protection), supporting services (such as nutrient cycles) and cultural services (such as spiritual and aesthetic values, recreation and cultural heritage) (MEA, 2005). Despite including cultural services, this paradigm has generally been applied to conserving natural resources and land use planning rather than managing cultural landscapes (Pleininger et al., 2014). However, recent work by the International Union of Conservation of Nature (IUCN) has integrated cultural and biological diversity. These approaches have now been supplemented by Nature-based Solutions[1]that highlight the role of cultural landscapes in comprehensive planning and the engagement of local communities. UNESCO has promoted the field of biocultural diversity by "stress(ing) the interrelated and co-evolving nature of biological and cultural systems, values, and practices" (Larsen & Wijesuriya, 2017, p. 144).[2] Others have studied agrobiodiversity conservation using cultural landscape knowledge (Zimmerer, 2014, 2019). Landscapes for People, Food, and Nature Initiative, co-organized by the United Nations Environment Programme, the United Nations Food and Agriculture Organization, and other international organizations, has sought to address rural landscape management that integrates food production, biodiversity and ecosystem conservation, and rural livelihoods within supportive institutional and policy contexts (Pleininger et al., 2014). The United Nations Sustainable Development Goals also address cultural landscapes.[3]

Resilience is "the ability of a social or ecological system to absorb disturbances while retaining the same basic structure and ways of functioning, the capacity for self-organization and the capacity to adapt to stress and change" (Aktürk & Dastgerdi, 2021, p. 2).

Traditional knowledge, or "traditional ecological knowledge," is a crucial feature of resilient systems in indigenous communities (Berkes et al., 2000; Pleininger & Bieling, 2013). Traditional knowledge refers to the ancient scientific and cultural knowledge passed down over many generations, which has produced local techniques supporting the sustainable management of ecosystems and cultural landscapes. This knowledge featured a deep environmental understanding and a certain social cohesion was vital to its success. The continuing landscapes in the Galilee and the Bedouin landscapes in the Negev reflect these traditions.

The resulting techniques were characterized by low resource consumption using locally sourced renewable resources while trying to derive the maximum output. Traditional knowledge is becoming increasingly scarce in our modern globalized world. However, many cultural landscapes rely upon these tools to continue to exist sustainably and provide additional models of continuous organic landscapes that have tradition embedded into their structure.

Indigenous and local knowledge has been addressed by the Intergovernmental Platform on Biodiversity and Ecosystem Services in order to incorporate such knowledge into environmental policy (Pleininger & Bieling, 2013). They define indigenous and local knowledge (ILK) as a "dynamic body of social-ecological knowledge, practice and belief, evolving by adaptive processes, rooted in territory, intergenerational and cultural transmission, about the relationship of living beings (including humans) with one

[1] According to IUCN, "nature-based solutions" leverage nature and the power of healthy ecosystems to protect people, optimize infrastructure and safeguard a stable and biodiverse future. https://www.iucn.org/our-work/nature-based-solutions.

[2] The relationship between cultural landscapes and sustainability is noted in paragraph in Paragraph 119 of the Operational Guidelines (2019) on sustainable use: "World Heritage properties may sustain biological and cultural diversity and provide ecosystem services and other benefits, which may contribute to environmental and cultural sustainability. Properties may support a variety of ongoing and proposed uses that are ecologically and culturally sustainable and which may enhance the quality of life and well-being of communities concerned. The State Party and its partners must ensure their use is equitable and fully respects the Outstanding Universal Value of the property. For some properties, human use would not be appropriate. Legislation, policies and strategies affecting World Heritage properties should ensure the protection of the Outstanding Universal Value, support the wider conservation of natural and cultural heritage, and promote and encourage the effective, inclusive and equitable participation of the communities, indigenous peoples and other stakeholders concerned with the property as necessary conditions to its sustainable protection, conservation, management and presentation".

[3] Article 9 acknowledges the world's natural and cultural diversity and recognizes that all cultures and civilizations can contribute to sustainable development. Sustainable Development Goal 11, which addresses the challenges of urbanism, references heritage most directly by calling on "strengthen[ing] efforts to protect and safeguard the world's cultural and natural heritage" to achieve the goal of making our "cities and human settlements inclusive, safe, resilient and sustainable."

another and with their environment" (Molnar & Berkes, 2018, p. 184).

Some indicators of the social-ecological resilience of landscape, particularly in the agricultural context, are the presence of community-based institutions; local knowledge of natural resources; use of traditional crop varieties; and the degree of heterogeneity and multifunctionality in the landscape (Plieninger & Bieling, 2012). Examples include the emergence of bottom-up partnerships that provide an alternative model to globalized agricultural practices and work toward safeguarding a regional quality of life, managing landscape change, and protecting natural resources. These multi-stakeholder rural networks integrate agriculture, nature conservation, and regional development based on collaborative management practices and local collectives (Plieninger & Bieling, 2012). These features are present in several of the case studies presented in this collection, especially the olive groves of the Galilee (Chap. 5) and the Megiddo Biosphere Region (Chap. 6).

1.1.4 Climate Change

Although many of the environmental issues already noted are bound up with the climate crisis, it is worth underscoring its singular impact as the most significant threat to cultural landscapes (Aktürk & Dastgerdi, 2021; Melnick, 2016). In 2005, The World Heritage Committee recognized climate change as an "emerging threat to the conservation of many cultural and natural sites" (Dastgerdi et al., 2019, p. 2). The Committee endorsed the report in 2007 (UNESCO, 2021), and subsequent additions and guidelines are continuing to be made. The impacts of climate change are both physical and cultural. Physical impacts include the increased frequency and intensity of severe weather events, extreme heat, sea-level rise and coastal flooding, degraded ecosystems and reduced biodiversity, drought and water scarcity, and erosion (Sesana et al., 2018). This is especially evident in Israel. Cultural impacts include changing livelihood patterns and, in extreme cases, the displacement of populations through forced migration, leading to the abandonment of property and the eventual loss of cultural memory and traditional knowledge required to maintain heritage properties (Dastgerdi et al., 2019).

There is an urgent need to address climate resilience in cultural landscapes. Some of the recommendations to date include strengthening regulatory frameworks to support collaboration and cooperation among all the local stakeholders (Aktürk & Dastgerdi, 2021), reinforcing monitoring and maintenance, and increasing risk-preparedness and adaptation measures. Access to guidelines and financial incentives is also needed to encourage organizations to commit to climate change adaptation (Sesana et al., 2018). Adaptation measures need to be proactive, not just reactive.

Recent research has also sought to strengthen the ties between heritage and landscape planning, stating that "the protection of the landscape cannot be based on simply conserving it, but rather in controlling the transformations that occur therein so that its identity values are not compromised" (López Sánchez et al., 2020, p. 2). From this perspective, it is not effective to argue for conservation versus development; cultural landscapes should be considered part of a larger sustainable landscape planning strategy (López Sánchez et al., 2020).

1.1.5 Contested Landscapes

Another significant framework for Israel's cultural landscapes is their multiple and often contested meaning. As geographer Donald Meinig forcefully argued in his essay "The Beholding Eye: Ten Versions of the Same Scene" (1979), the cultural landscape can hold competing sets of meaning for different individuals. Landscapes are often contested spaces (Bender & Winer, 2001), subject to conflicting interpretations and meanings. Appadurai articulated the inherently political character of heritage sites and cultural landscapes: "collectively held, publicly expressed and ideologically charged... versions of the past... are likely to vary within the groups that form a society... discourse concerning the past between social groups is an aspect of politics, involving competition, opposition and debate" (cited in Silverman, 2011, p. 24). The landscapes of Israel are a prime example.

The political character of cultural landscapes implies that power relations are often critical to shaping heritage and identity (Whelan, 2016). Based on their study of Masada, the ancient Israeli site of heroism and resistance and a site of complex, competing narratives, Bruner and Gorfain (1984) developed the argument that identities are not fixed or timeless but under constant construction. This sense of "constructedness" resonated with Hobsbawm's notion of the invention of tradition, referring to the selective creation of the past (1983). This idea would become central to discussions of heritage and cultural landscapes in the 1990s, especially concerning debates over authenticity (Lowenthal, 2015; Silverman, 2011). Jerusalem's Old City is another significant example of a contested heritage site, particularly regarding the recreation of the destroyed Jewish Quarter in the Old City of Jerusalem (Bar & Rubin, 2011; Ricca, 2003).

At times, a cultural landscape may be associated with different communities that exist side by side and, as a result, may reflect multiple conflicting cultural, religious, or nationalistic narratives. Each community may selectively preserve specific landscape elements that strengthen its narrative at the expense of features that do not support the narrative. The issue of power is particularly apparent in contested landscapes. Whoever has greater power can shape the cultural landscape in a way that privileges their own history and values (Department of Environment and Climate Change NSW, 2008). However, if there is mutual respect rather than a culture of delegitimization, different communities may coexist despite their clearly conflicting narratives. This is best seen in the 1852 *firman* (decree) on the Declaration of the Status Quo in the Holy Places in Jerusalem and Bethlehem (Breger et al., 2009). Thus, the concept of contested landscapes can be nuanced, inviting a range of responses (Amit-Cohen, 2005). It may take the form of activist, community-curated content and decision-making (Kryder-Reid, 2018). Also known as Critical Heritage, this approach encompasses a variety of bottom-up methodologies and strategies for empowering diverse stakeholders, "deploying tangible and intangible heritage to address historical and systemic inequalities as a social activist strategy" (Kryder-Reid, 2018). These strategies are essential for creating new forms of interpretation and managing contested landscapes.

In Israel, cultural landscapes are often highly contested, adding complexity while also presenting an important challenge and opportunity. The Burra Charter (Australia ICOMOS, 2013) recommended that the coexistence of cultural values should always be recognized, respected, and encouraged and that all aspects of cultural significance should be expressed. Article 6 of the Nara Document (ICOMOS, 1994, 2014) states that "cultural heritage diversity exists in time and space and demands respect for other cultures and all aspects of their belief systems. In cases where cultural values appear to be in conflict, respect for cultural diversity demands acknowledgment of the legitimacy of the cultural values of all parties" (ICOMOS, 1994). The ICOMOS Charter for the Interpretation and Presentation of Cultural Heritage Sites (known as the "Ename Charter") also addresses these issues (ICOMOS, 2008).

The cultural landscape approach to heritage management recognizes the value of diverse associations and meanings embedded in the landscape. Therein lies one of the great challenges of negotiating contested and multifaceted meanings and preserving the landscape in a manner that supports the reading of potentially divergent messages.

1.1.6 Community Engagement and Livelihood-Based Approaches

The pluralistic approach discussed above requires a more participatory design and decision-making process. In recent decades, community engagement practices have become ever more critical in safeguarding and managing cultural landscapes. These include aspects of interpretation, educational programs, and broad outreach to stakeholders. Globally, this has included collaborating and empowering indigenous communities, especially to preserve local knowledge systems central to sustainable agriculture. A fundamental shift has taken place from expert-driven to participatory heritage.

A second theme related to community engagement concerns livelihood-based approaches. IUCN's Commission has recognized this approach to environmental, economic, and social policy, known as "Human Well-being and Sustainable Livelihoods."[4] The HWSL Theme is based on the belief that "conservation efforts that recognize indigenous Peoples' and local communities (IPLCs) ways of knowing, being and doing, and organize around those values, practices and institutions, are the most effective at conserving nature." While the Operational Guidelines of the World Heritage Convention encourage consultation, they do not directly empower local communities. The Ramat Menashe stakeholders have adopted the more proactive outlined in the Man and Biosphere (MaB) model (see Chap. 6).

Livelihood-based approaches also strengthen agrobiodiversity by including non-traditional groups, such as indigenous peoples or small producers (Zimmerer et al., 2019). Israel is far behind in many facets of community engagement in the planning process.

References

Aktürk, G., & Dastgerdi, A. S. (2021). Cultural landscapes under the threat of climate change: A systematic study of barriers to resilience. *Sustainability, 13*, 9974.

AlSayyad, N. (2001). Global norms and urban forms in the age of tourism: manufacturing heritage, consuming tradition. In *Consuming tradition, manufacturing heritage: Global norms and urban forms in the age of tourism* (pp. 1–33).

Amit-Cohen, I. (2005). Synergy between urban planning, conservation of the cultural built heritage and functional changes in the old urban center—The case of Tel Aviv. *Land Use Policy, 22*(4), 291–300.

Amit-Cohen, I., & Amram, G. (2002). Changes in usage in the Old Center of Tel Aviv and attitudes towards conserving built heritage. *Research of the Geography of the Land of Israel, 16*, 289–316.

Ashworth, G. J. (1994). From history to heritage–from heritage to identity. In *Building a new heritage: Tourism, culture and identity in the new Europe* (pp. 13–30). Routledge.

Ashworth, G., & Dehan, T. Z. (1986). *Uses and users of the tourist historic city: an evolutionary model in Norwich*. Serieveld Studies: Geografisch Institute RIJK Suniversitiet, Groningen.

Ashworth G.J., & Larkham P. J. (1994). A heritage for Europe: The need, the task, the contribution?. In G. J. Ashworth & P. J. Larkham (Eds). *Building a new heritage, tourism, culture and identity in the New Europe* (pp. 1–9). London and New York.

[4] https://www.iucn.org/commissions/commission-environmental-economic-and-social-policy/our-work/human-wellbeing-and-sustainable-livelihoods.

Australia, ICOMOS. (2013). *The Burra Charter: The Australia ICOMOS charter for places of cultural significance*. Burra: ICOMOS.

Azaryahu, M. (1995) State Worship-Independence celebrations and the memorialization of the fallen in Israel 1948–1956. The Heritage Center for Ben Gurion.

Bar, D., & Rubin, R. (2011). The Jewish Quarter after 1967: A case study on the creation of an ideological-cultural landscape in Jerusalem's old city. *Journal of Urban History, 37*(5), 775–792.

Barnes, T.J., & Duncan, J. S. (1992). Introduction: Writing worlds. In T. J. Barnes & J. S. Duncan (Eds), *Writing Worlds*. London: Routledge.

Bender, B., & Winer, M. (2001). *Contested Landscapes: Movement*. Exile and Place.

Berkes, F., Colding, J., & Folke, C. (2000). Rediscovery of traditional ecological knowledge as adaptive management. *Ecological Applications, 10*(5), 1251–1262.

Breger, Y. R. M. J., Reiter, Y., & Hammer, L. (2009). *Holy places in the Israeli-Palestinian conflict*. Taylor & Francis.

Bruner, E. M., & Gorfain, P. (1984). Dialogic narration and the paradoxes of Masada. In *Text, play and story: The construction and reconstruction of self and society* (pp. 56–79).

Cosgrove, D., & Jackson, P. (1987). New directions in cultural geography. *Area*, 95–101.

Daniels, S. (1993) *Fields of vision: Landscape imagery and national identity in England and in United States*. Princeton.

Dastgerdi, A. S., Sargolini, M., & Pierantoni, I. (2019). Climate change challenges to existing cultural heritage policy. *Sustainability, 11* (19), 5227.

Duncan, J. (1990). *The city as text: The politics of landscape interpretation in the Kandyan Kingdom*. Cambridge University Press.

Duncan, J. S., & Duncan, N. G. (2001). The aestheticization of the politics of landscape preservation. *Annals of the Association of American Geographers, 91*(2), 387–409.

Elkins, T. H. (1989). Human and regional geography in the German-speaking lands in the first forty years of the Twentieth Century. In J. Nicholas Entrikin & S. D. Brunn (eds.), *Reflections on Richard Hartshorne's The nature of geography, occasional publications of the Association of the American Geographers*, Washington DC (pp. 17–34).

Everson, P., & Williamson, T. (1998). *The archaeology of landscape: Studies presented to Christopher Taylor*. Manchester University Press.

Featherstone, M. (2007). Consumer culture and postmodernism. *Consumer Culture and Postmodernism*, 1–232.

Groth, P. E., & Bressi, T. W. (Eds.). (1997). *Understanding ordinary landscapes*. Yale University Press.

Hewison, R. (1987). *The heritage industry: Britain in a climate of decline*. Methuen.

Hobsbawm, E. J., & Terence, O. (1983). *The invention of tradition*. Cambridge.

Hooke, D. (1998). *The landscape of Anglo-Saxon England*. Burns & Oates.

ICOMOS (1994, 2014). *The Nara document on authenticity*. Nara: ICOMOS - International Council on Monuments and sites. Retrieved June 23, 2020, from https://www.icomos.org/charters/nara-e.pdf.

ICOMOS (1999). *Charter for the conservation of places of cultural significance* (The Burra charter).

ICOMOS (2008) ICOMOS charter for the interpretation and presentation of cultural heritage sites, *ICOMOS*, Canada. Retrieved Oct 28, 2022, from http://icip.icomos.org/downloads/ICOMOS_Interpretation_Charter_ENG_04_10_08.pdf.

Jackson, J. B. (1986). *Discovering the vernacular landscape*. Yale University Press.

Jackson, J. B. (1994). *A sense of place, a sense of time*. Yale University Press.

Kryder-Reid, E. (2018). Introduction: Tools for a critical heritage. *International journal of heritage studies* (Vol. 24, pp. 691–693). Taylor & Francis.

Larsen, P. B., & Wijesuriya, G. (2017). Nature–Culture interlinkages in world heritage: Bridging the gap. In *The George Wright forum* (Vol. 34, pp. 142–153).

López Sánchez, M., Tejedor Cabrera, A., & Linares Gómez Del Pulgar, M. (2020). Guidelines from the heritage field for the integration of landscape and heritage planning: A systematic literature review. *Landscape and Urban Planning, 204*, 103931.

Lowenthal, D. (2015). *The past is a foreign country- Revisited*. Cambridge.

Meinig, D. W. (1979). The beholding eye. In D. W. Meinig (ed.), *The interpretation of ordinary landscapes*. Oxford University Press.

Melnick, R. Z. (2016). Deciphering cultural landscape heritage in the time of climate change. *Landscape Journal, 35*(2), 287–302.

Millennium Ecosystem Assessment. (2005). *Ecosystems and Human Well-being: Synthesis Report*. Washington, DC: Island Press.

Molnár, Z., & Berkes, F. (2018). Role of traditional ecological knowledge in linking cultural and natural capital in cultural landscapes. In M. L. Paracchini, P. C. Zingari, & C. Blasi (Eds.), *Reconnecting natural and cultural capital: Contributions from science and policy* (pp. 183–193).

Muir, R. (1999). *Approaches to landscape*. Springer.

Osborne, B. S. (1998). Constructing landscapes of power: The George Etienne Cartier monument, Montreal. *Journal of Historical Geography, 24*(4), 431–458.

Plieninger, T., & Bieling, C. (2012). *Resilience and the cultural landscape: Understanding and managing change in human-shaped environments*. Cambridge University Press.

Plieninger, T., & Bieling, C. (2013). Resilience-based perspectives to guiding high-nature-value farmland through socioeconomic change. *Ecology and Society, 18*(4), 20.

Plieninger, T., Van der Horst, D., Schleyer, C., & Bieling, C. (2014). Sustaining ecosystem services in cultural landscapes. *Ecology and Society*, *19*(2), 59.

Ricca, S. (2003). *The Jewish Quarter of Jerusalem: analysis of the destruction and reconstruction since 1967*. University of Exeter.

Sauer, C. (1925). *The morphology of landscape*. University of California Press.

Schein, R. H. (1997). The place of landscape: A conceptual framework for interpreting an American scene. *Annals of the Association of American Geographers, 87*(4), 660–680.

Sesana, E., Gagnon, A. S., Bertolin, C., & Hughes, J. (2018). Adapting cultural heritage to climate change risks: Perspectives of cultural heritage experts in Europe. *Geosciences, 8*(8), 305.

Shamir, A. (1996). *Memory and Memorialization: Israeli society and the design of Memorial Landscapes*. (Hebrew).

Silverman, H. (2011). Contested cultural heritage: A selective historiography. In *Contested cultural heritage* (pp. 1–49). Springer.

UNESCO. (2021). Operational guidelines for the implementation of the world heritage convention. UNESCO World Heritage Centre. https://whc.unesco.org/en/guidelines/.

Whelan, Y. (2016). *Heritage, memory and the politics of identity: New perspectives on the cultural landscape*. Routledge.

Wylie, J. (2007). *Landscape*. Routledge.

Zimmerer, K. S. (2014). Conserving agrobiodiversity amid global change, migration, and nontraditional livelihood networks: the dynamic uses of cultural landscape knowledge. *Ecology and Society*, *19*(2), 1.

Zimmerer, K. S., de Haan, S., Jones, A. D., Creed-Kanashiro, H., Tello, M., Carrasco, M., & Olivencia, Y. J. (2019). The biodiversity of food and agriculture (Agrobiodiversity) in the anthropocene: Research advances and conceptual framework. *Anthropocene*, *25*, 100192.

2 Mechanisms and Tools

2.1 The World Heritage Convention

The proposed cultural landscapes in Israel are classified according to the categories defined in the Operational Guidelines of the World Heritage Convention (UNESCO, 2021). In 1972, UNESCO recognized the "combined works of nature and man" with the adoption of the *Convention Concerning the Protection of the World Cultural and Natural Heritage*. The Convention aimed to protect cultural and natural heritage with "Outstanding Universal Value," a term that refers to outstanding cultural or natural importance that transcends geopolitical borders. These sites have "significant interaction between people and the natural environment" (UNESCO, 2021).

The World Heritage Convention was the first international framework to address integrated natural and cultural heritage; it "established a unique international instrument recognizing and protecting both cultural and natural heritage of Outstanding Universal Value" (Rossler, 2006, p. 1). However, the divide between the two realms remained (Fowler, 2002). Each was addressed in separate articles of the Convention, and there was little understanding of the complex interplay between the two (Brumann & Gfeller, 2022; Gfeller, 2013; Larsen & Wijesuriya, 2017; Taylor & Lennon, 2012). Indeed, the criteria for inscription were in two separate lists, six for culture and four for nature. These two lists were combined in 2005 to form a single set of ten criteria.

Most sites were featured for their cultural or their natural significance, and these were evaluated by distinct organizations. The International Council on Monuments and Sites (ICOMOS) reviewed sites with cultural significance while the International Union for the Conservation of Nature (IUCN) was responsible for sites of natural significance. This continued to be the case even when "mixed properties" began to be inscribed in the World Heritage List, such as Tikal National Park in Guatemala (1979), which was listed under both cultural and natural criteria for its archaeological and bio-ecological features (Brumann & Gfeller, 2022).

The divide between natural and cultural heritage was the subject of debate throughout the 1980s, as was the Eurocentric focus of the conception of heritage (Brumann & Gfeller, 2022; Gfeller, 2013). As a result of the rising awareness of cultural landscapes on the part of government agencies, NGO's and civil society, a breakthrough occurred in 1992 when the World Heritage Committee adopted categories of World Heritage cultural landscape and revised the cultural criteria for inscribing properties on the World Heritage List. Cultural heritage was defined in 1992 in the Operational Guidelines based on Article 1 of the *Convention* (UNESCO, 2003a, b), using three main categories: *created or designed landscapes, organically evolved landscapes, and associative landscapes.* These categories fundamentally redefined the approach to heritage; heritage sites were no longer viewed in isolation but were now evaluated contextually with an emphasis on local communities' relationship to their environment (Taylor & Lennon, 2012). From their earlier, narrow focus on archaeological sites, individual monuments and historical architectural ensembles—what Taylor characterized as "elite connoisseurship" (2016, p. 211)—World Heritage inscriptions now reflected everyday living landscapes of "peoples' livelihoods, identities and belief systems all over the world" (Larsen & Wijesuriya, 2017, p. 148). "The concept of 'cultural landscapes' as World Heritage sites, then, embraces ideas of belonging, outstanding, significance, locality, meaning, value and singularity of place" (Fowler, 2002, p. 19).

Adopting cultural landscape categories could now address a diverse array of landscape types. Specifically, "associative landscapes" stressed the immaterial and intangible aspects of culture, religion, and spirituality. According to UNESCO, the associative landscape category was "particularly crucial in the recognition of intangible values and the heritage of local communities and indigenous people" (Gfeller, 2013). The formal recognition of intangible heritage occurred in 2003, with the approval by the UNESCO of the *Convention for the Safeguarding of the Intangible Cultural Heritage*. The criterion for intangible cultural

A. Sar Shalom et al., *Cultural Landscapes of Israel*,
https://doi.org/10.1007/978-3-031-33685-0_2

heritage (Criterion vi) embraced a broad range of cultural practices, knowledge and skills, including: "oral traditions…; social practices, rituals and festive events; knowledge and practices concerning nature and the universe and traditional craftsmanship" (Mitchell et al., 2009, p. 27).

These changes also helped bridge the divide between nature and culture by including diverse landscape types. (Brumann & Gfeller, 2022; Gfeller, 2013).[1] However, as Larsen and Wijesuriya (2017) and others have argued, while the recognition of cultural landscape has allowed for new linkages between nature and culture, it has also effectively maintained their separation. "Mixed sites," inscribed for both natural and cultural values, make up only 3% of the World Heritage List (Larsen & Wijesuriya, 2017, p. 148) due to the complex institutional challenges involved, from their nomination process to their evaluation and management.

Another problematic aspect of the concept of protection is the management of change. As Sánchez et al. have written, "The protection of the landscape cannot be based on simply conserving it, but rather in controlling the transformations that occur therein so that its identity values are not compromised" (2020, p. 2).

2.2 Other Mechanisms for the Protection of Landscapes

2.2.1 Global Mechanisms

An expanding list of conventions and charters has been developed by the world organizations of UNESCO, ICOMOS, and IUCN. These provide mechanisms to identify and protect an increasingly varied set of cultural landscape typologies. While conventions establish legally binding norms, charters do not carry legal authority but serve as guidelines for professional activity.[2] UNESCO Conventions, for example, have addressed intangible cultural heritage as well as the protection of the diversity of cultural expression.[3] UNESCO Recommendations, such as the Recommendation concerning the Safeguarding of Beauty and Character of Landscapes and Sites, 1962, and the Recommendation on the Historic Urban Landscape, 2011, together with ICOMOS Charters, have addressed typologies such as the conservation of historic towns and urban areas, archaeological heritage, underwater cultural heritage, tourism, built vernacular heritage, cultural routes, rural landscapes, and military heritage.[4]

In addition to the World Heritage listing, two other significant UNESCO designations provide holistic frameworks for managing biological and cultural diversity: biosphere reserves (1971) and global geoparks (2001). "While Biosphere Reserves focus on the harmonised management of biological and cultural diversity and World Heritage Sites promote the conservation of natural and cultural sites of Outstanding Universal Value, UNESCO Global Geoparks give international recognition for sites that promote the importance and significance of protecting the Earth's geodiversity through actively engaging with the local communities."[5] A third framework is the peace park that may be designated as an IUCN WCPA Transboundary Conservation Area[6] and is "dedicated to the promotion, celebration and/or commemoration of peace and cooperation" (Erg et al., 2015).

Several international conventions focus on biodiversity issues.[7] In 2002, the Globally Important Agricultural Heritage Systems (GIAHS) initiative was established by the UN Food and Agriculture Organization (FAO) as means of safeguarding biodiversity at the genetic and the landscape level, that is, to conserve indigenous varieties and species as well as protect supporting ecosystem services such as clean water and food security. The program stresses "dynamic conservation" by supporting "agricultural systems that sustain and conserve our biodiversity and genetic resources for food and agriculture, rural livelihoods, knowledge systems, cultures and remarkable landscapes."[8]

The 2030 Sustainable Development Goals also address cultural landscapes. Article 9 acknowledges the natural and

[1] See *Connecting Practice Project* (2015), a joint project of IUCN and ICOMOS that addressed linking nature and culture. See https://www.icomos.org/en/178-english-categories/news/3580-just-released-final-report-of-connecting-practice-project.

[2] (https://www.getty.edu/conservation/publications_resources/newsletters/19_2/).

[3] See UNESCO Convention of the Safeguarding of the Intangible Cultural Heritage (2003) and UNESCO Convention on the Protection and Promotion of the Diversity of Cultural Expressions (2005).

[4] See ICOMOS Charter for the Conservation of Historic Towns and Urban Areas (1987); ICOMOS Charter for the Protection and Management of Archaeological Heritage (1990); ICOMOS Charter on the Protection and Management of the Underwater Cultural Heritage (1996); ICOMOS International Cultural Tourism Charter—Managing Tourism at Places of Heritage Significance (1999); ICOMOS Charter on the Built Vernacular Heritage (1999); Joint ICOMOS—TICCIH Principles for the Conservation of Industrial Heritage Sites, Structures, Areas and Landscapes (2011); ICOMOS-IFLA Principles Concerning Rural Landscapes as Heritage (2017).

[5] (https://en.unesco.org/global-geoparks).

[6] IUCN World Commission on Protected Areas—Transboundary Conservation Specialist Group; https://www.iucn.org/our-union/commissions/group/iucn-wcpa-transboundary-conservation-specialist-group.

[7] The Convention on Biological Diversity entered into force in 1993. (https://www.un.org/en/observances/biological-diversity-day/convention#:~:text=The%20Convention%20on%20Biological%20Diversity,been%20ratified%20by%20196%20nations).

[8] (https://www.fao.org/publications/card/en/c/09cbda9a-73bf-4a28-bec3-dcfcf32621a8/).

cultural diversity of the world and recognizes that all cultures and civilizations can contribute to sustainable development. Paragraph 36 of the Declaration's Preamble states, "We pledge to foster intercultural understanding, tolerance, mutual respect and an ethic of global citizenship and shared responsibility. We acknowledge the natural and cultural diversity of the world and recognize that all cultures and civilizations can contribute to, and are crucial enablers of, sustainable development" (United Nations, 2015). The most direct reference to heritage is found in Goal 11, which addresses the challenges of urbanism. Heritage is highlighted in a single target 11.4: "strengthen[ing] efforts to protect and safeguard the world's cultural and natural heritage" to achieve the goal of making our "cities and human settlements inclusive, safe, resilient and sustainable" (United Nations, 2015, p. 22). Cultural landscapes can support sustainability around human settlements and fulfill the role of developing sustainable tourism, creating jobs and promoting local culture and products (Goal 12b).

Many countries have acknowledged the goal of sustainable development on global, regional, and national levels. In such cases, a holistic approach to planning and development is encouraged, integrating strict guidelines for using the landscape and its resources. In this way, landscapes reflect society's perceptions, traditions, and development and maintain their outstanding natural and cultural heritage to preserve them for the use and pleasure of future generations (Ewald, 2001).

As of 2023, 121 properties with six transboundary properties (1 delisted property) on the World Heritage List have been included as cultural landscapes.[9]

Sites are compiled into an inventory, known as a "Tentative List" of national properties "situated on its territory which each State Party considers suitable for nomination to the World Heritage List."[10] In 2019, the World Heritage Committee endorsed a two-phase nomination process: a preliminary assessment must be completed in the first phase prior to the site's nomination in the second phase. In the first phase, a process of study and documentation follows to determine Outstanding Universal Value as well as the effective and inclusive participation of local communities and stakeholders, and a discussion regarding the implications of conservation methods, including a management plan or system, maintenance, and monitoring programs.[11] While there is widespread agreement regarding the inscription process, the criteria for selecting cultural landscapes and ranking their significance have proved far more challenging. Cultural landscapes relate to specific socio-geographic contexts. As such, some States Parties tend to favor places of national significance, while others prefer to focus on particular historical periods while disregarding others. The goal of cultural diversity demands that a more balanced approach be taken to identify these narratives.

2.2.2 Regional and National Mechanisms

Several mechanisms have been created over the past forty years to conserve and protect heritage components that have a bearing on cultural landscapes. The European Landscape Convention, approved in 2000, provides a more broad-based and radical definition of the role of the landscape—it covers natural, rural, urban, and peri-urban areas and includes outstanding as well as everyday landscapes. The Convention defines landscape protection as actions to conserve and maintain a landscape's significant or characteristic features, justified by its heritage value and derived from its natural configuration or human activity. It emphasizes landscape planning as a means of providing a framework to enhance, restore or design landscapes. From the perspective of sustainable development, management is viewed as a means of ensuring the regular maintenance of the landscape to guide and harmonize changes brought about by social, economic, and environmental processes: "...it should be remembered that the aim is not the preservation or "freezing" of the landscape at a particular point in its lengthy evolution. Landscapes have always changed and will continue to change, both through natural processes and through human action. The aim should be to manage future changes in a way which recognizes the great diversity and the quality of the landscapes that we inherit and which seeks to preserve, or even enhance, that diversity and quality instead of allowing them to decline." (CoE, 2000). It has been noted that in contrast to the reactive stance of the World Heritage Convention, the definition of landscape proposed by the European Landscape Convention (2000) offers a proactive approach (Aktürk & Dastgerdi, 2021). It emphasizes landscape planning as a framework to enhance, restore or design landscapes and stresses the need for an integrated action between all sectors, including heritage.

Two more recent initiatives, the proposed International Landscape Convention and the Latin American Landscape Initiative, were inspired by the European Landscape Convention and sought to expand its application. During the 1990s, many countries developed their definitions of cultural landscapes parallel to those approved by the World Heritage Committee. They prepared landscape inventories based on their own definitions adding site-specific categories and criteria. Some countries did more than add new definitions;

[9] (https://whc.unesco.org/en/culturallandscape/).

[10] Operational Guidelines; Paragraph 62, Tentative Lists—Procedure and Format.

[11] Operational Guidelines; Paragraph 123, Participation in the nomination process.

they went as far as to provide detailed criteria for classifying the landscapes, adding the means for their registration, and guidance for cultural landscape studies and preservation interventions. At the national level, the United States National Parks Service undertook pioneering work on cultural landscapes, as did English Heritage in the United Kingdom. While both agencies have created tools to manage the landscapes under their purview, their approaches differ significantly. Each agency has developed its distinctive set of tools adapted to suit local conditions and serve specific agendas. These national mechanisms have developed over nearly half a century and illustrate these frameworks' evolution from theoretical and operational perspectives.

Public participation has gained support since the approval of the European Landscape Convention in 2000, which required that all participating countries involve the public in every aspect of the inscription process (Selman, 2004). Such cooperation is intended to guarantee that the interests of the general public are protected. In his examination of the inscription process of cultural landscapes and World Heritage properties, Coeterier (2002) demonstrated that classification by specialists is insufficient. He argued that while the process focuses on landscape components, physical structure, and design, the site must also appeal to the public to whom the property or the site belongs.

Linehan and Gross (1998) identified three distinct approaches to the landscape inscription process: ecological, economic, and sociocultural. The ecological approach concerns the possible damage or disappearance of landscapes due to over-development. The economic approach emphasizes the costs of preservation versus the profitability of development, while the sociocultural approach emphasizes society's needs and attitudes toward the landscape. Linehan and Gross argued that the third approach is preferred; while planning processes must consider ecological and economic limitations, these parameters alone are insufficient. To promote the sustainable preservation of cultural landscapes for future generations, it is vital to generate awareness among members of the local community and to examine the community's preferences and perceptions of its landscapes.

2.3 Directions for Israel

Which properties are to be deemed worthy of conservation in Israel? How should they be ranked, and what kinds of limits should be placed on development? Just as the United States and the United Kingdom have developed their own definitions of cultural landscapes, Israel, too, has adapted global approaches to local conditions and stakeholder needs. The international tools and mechanisms discussed above should assist local planners and stakeholders in understanding the range of policies developed at the national and international levels to define, conserve and manage the cultural landscape.

References

Aktürk, G., & Dastgerdi, A. S. (2021). Cultural landscapes under the threat of climate change: A systematic study of barriers to resilience. *Sustainability, 13*, 9974.

Brumann, C., & Gfeller, A. É. (2022). Cultural landscapes and the UNESCO World Heritage List: Perpetuating European dominance. *International Journal of Heritage Studies, 28*(2), 147–162.

Coeterier, J. F. (2002). Lay people's evaluation of historic sites. *Landscape and Urban Planning, 59*(2), 111–123.

Council of Europe. (2000). European landscape convention–explanatory report. Council of Europe, Strasbourg [on-line]. Retrieved July 25, 2013, from http://conventions.coe.int/Treaty/EN/Reports/Html/176.htm.

Erg, B., Groves, C., McKinney, M., Michel, T. R., Phillips, A., Schoon, M. L., et al. (2015). Transboundary conservation: A systematic and integrated approach. In *Best practice protected area guidelines series no. 23*.

Ewald, K. C. (2001). The neglect of aesthetics in landscape planning in Switzerland. *Landscape and Urban Planning, 54*(1–4), 255–266.

Fowler, P. (2002). World heritage cultural landscapes, 1992–2002: A review and prospect. In *Cultural landscapes: The challenges of conservation* (p. 16).

Gfeller, A. E. (2013). Negotiating the meaning of global heritage: 'cultural landscapes' in the UNESCO World Heritage Convention, 1972–92. *Journal of Global History, 8*(3), 483–503.

Larsen, P. B., & Wijesuriya, G. (2017). Nature–culture interlinkages in World Heritage: Bridging the gap. In *The George Wright Forum* (Vol. 34, pp. 142–153). JSTOR.

Linehan, J. R., & Gross, M. (1998). Back to the future, back to basics: The social ecology of landscapes and the future of landscape planning. *Landscape and Urban Planning, 42*(2–4), 207–223.

Mitchell, N., Rossler, M., & Tricaud, P.-M. (2009). *World Heritage paper no. 26. World Heritage cultural landscapes. A handbook for conservation and management*. 4/2/UNESCO/Cult/09/E.

Rossler, M. (2006). *World Heritage cultural landscapes*. Taylor & Francis.

Sánchez, M. L., Cabrera, A. T., & Del Pulgar, M. L. G. (2020). Guidelines from the heritage field for the integration of landscape and heritage planning: A systematic literature review. *Landscape and Urban Planning, 204*, 103931.

Selman, P. (2004). Community participation in the planning and management of cultural landscapes. *Journal of Environmental Planning and Management, 47*(3), 365–392.

Taylor, K. (2016). Cultural landscape meanings and values. In *Research in landscape architecture* (pp. 211–234). Routledge.

Taylor, K., & Lennon, J. (2012). *Managing cultural landscapes*. Routledge.

UN General Assembly, *Transforming our world: The 2030 agenda for sustainable development*, 21 October 2015, A/RES/70/1. Retrieved Aug 30, 2022, from https://www.refworld.org/docid/57b6e3e44.html.

UNESCO. (2003a). *Cultural landscapes: The challenges of conservation*. UNESCO.

UNESCO. (2003b). *Convention for the safeguarding of intangible cultural heritage*. UNESCO.

UNESCO (2021). Operational guidelines for the implementation of the world heritage convention, UNESCO World Heritage Centre, Paris. https://whc.unesco.org/en/guidelines/.

UNESCO Global geoparks. https://en.unesco.org/global-geoparks.

United Nations (2015). Transforming our world: The 2030 agenda for sustainable development.

3 Cultural Landscapes in Israel

Israeli cultural landscapes reflect a long and complex settlement history. From the ancient sites associated with some of the earliest agricultural practices in the world to those intimately bound up with the world's three major monotheistic traditions—Judaism, Christianity, and Islam—these landscapes hold a distinctive place in the Western cultural imagination.

Israel's unique geomorphological range has played a significant role in forming local cultural landscapes over the past millennia. One of the most dramatic aspects of the Israeli landscape is its rich diversity, given its relatively small size. Situated on the eastern coast of the Mediterranean Sea, the country has been a land bridge between continents with the Via Maris, a meeting point of climate zones between verdant areas and arid deserts. It is the place where cultures originated and flourished over successive eras. Its vital location on the continental crossroads has made it a thriving cultural and economic center that has always been open to diverse external cultural influences.

Israel also contains many associative and commemorative landscapes representing a wide range of stories, lifestyles, technologies, events, and characters. Relicts of landscapes have been identified that signify periods, traditions, construction styles, and technologies that no longer exist, as well as practices that have been maintained by adapting to changes over time. Some landscapes reflect traditions that have been renewed after years of being dormant.

Like other countries, Israel has added local characterizations to the 1992 definitions of cultural landscapes laid out in the Operational Guidelines of the Convention. We present several characteristics specific to the Israel geographic-historic-cultural context that are given special attention and focus.

3.1 The Geomorphology and Biomes of Israel

The total area of Israel is 20,770 km^2. It extends 424 km from north to south, while at its narrowest point, it is a mere 15 km wide from east to west. Its geomorphology is influenced by its location between two tectonic regions and its varied climatic zones. It is located at the southernmost limit of the typical Euro-Mediterranean climate. The coastal region of the Mediterranean features rich soils, while the temperate region in the north is the product of geologic karst processes. The Negev desert in the south comprises half of the country's land mass, with aeolian weathering processes shaping its topography along with the erosive force of flash floods that surge through low-lying areas (Nir, 1970). The southernmost tip of the country reaches the coast of the Gulf of Aqaba.

Israel's geologic profile is also dramatic. The country is divided by a mountain range with a north–south axis, starting from the Carmel Mountain range that plunges toward the Jordan Rift Valley in the east, where the Dead Sea is located at the lowest point on earth, currently at −417 m. The mountain regions of the Upper and Lower Galilee are separated from this range by the Jezreel Valley to the north.

Four biomes exist in the narrow Mediterranean and Dead Sea corridor: Mediterranean, Irano-Turanian, Saharo-Arabian, and Sudanic (Fig. 3.1). The IUCN defines biomes as "a major portion of the living environment of a particular region… characterized by its distinctive vegetation and maintained largely by local climatic conditions" (IUCN, 2022). Rainfall varies significantly across the country, with annual precipitation ranging from 25 mm in the

A. Sar Shalom et al., *Cultural Landscapes of Israel*,
https://doi.org/10.1007/978-3-031-33685-0_3

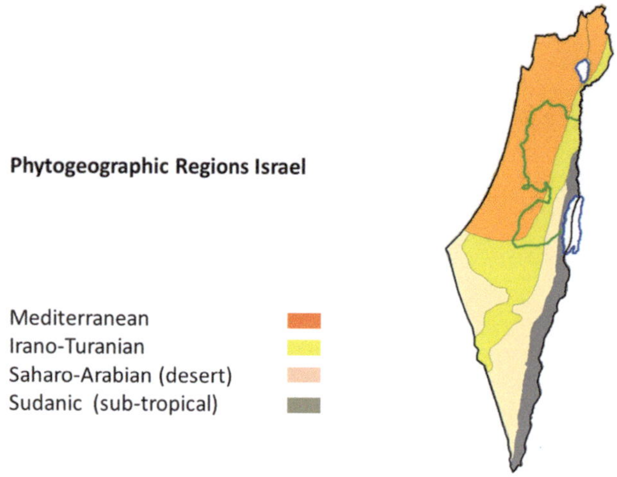

Fig. 3.1 Phytogeographic regions of Israel

southernmost tip to 1000 mm in the northern mountains (Ziv et al., 2014). Israel's strategic location as a land bridge is particularly apparent in biogeographical processes and plant and wildlife distribution patterns.

The Mediterranean biome is located along Israel's western coast and extends to the northeastern edge of the country. These landscapes feature forests, woodlands, and scrub, characterized by dry summers and wet winters. The Irano-Turanian biome is located in the east, north of the Sea of Galilee. It extends in a narrow band southward before sweeping across the Northern Negev Desert in a wide arc, nearly reaching the Gaza Strip. An additional extension of the Irano-Turanian biome reaches the Sinai peninsula, surrounded on both sides by the Saharo-Arabian biome, which spans much of Southern Israel and wraps around the Irano-Turanian biome further inland. The Sudanic biome forms a narrow ribbon near the Red Sea coast and extends northward to the low-lying Arava in the Jordan Rift Valley.

These factors contribute to the local landscapes, both natural and human-made: natural flora and fauna and historic relics of cultural and religious significance. Israel also has many patterns of transitions between settled land and wilderness. The transition can be sudden or gradual, permanent or dynamic. There are often sharp transitions from areas of permanent residence to those that are transient, from cultivated and irrigated fields to pasture lands irrigated by rainfall and some areas of transhumance. Affected by climate change and technological development, these transitions may change from year to year due to the encounter between human activity and natural processes.

3.2 Cultural Landscape Narratives

A steering committee comprising representatives from the Israel World Heritage Committee and the Israel Nature and Parks Authority was established in 2004 to trace the key narratives embedded in Israel's landscapes. It soon became apparent that this process required a diverse team of experts from the fields of history, archaeology, geography, architecture, and planning to explore the multilayered and multifaceted landscape. The result was a deliberately open-ended database that could allow for the future development of a more complete, dynamic narrative of the country's cultural heritage and landscapes.

The database spans some 5000 years, from the Bronze Age to the twentieth century. The biblical landscape includes the geographical settings of Old Testament texts such as Elijah and the Prophets of Ba'al, David and Goliath, and Saul and the Philistines, which are etched into religious and national narratives. Similarly, as described in the New Testament, Christian sites in the Galilee have been recognized as important associative cultural landscapes with a sacred character. Other significant historical events are recorded and narrated in place, such as the Bar Kokhba rebellion of the Jews against the Romans at the Land of the Caves, and the epic battle between the Crusaders and the Ayyubid army of Saladin at the Horns of Hattin, which both contain historic physical evidence. Prehistoric, Canaanite, Israelite, Christian, Nabataean, Byzantine, Islamic, Baha'i, Circassian, Ottoman, Templar, and British mandatory eras are all represented in the cultural landscape narratives.

Four main themes can be identified: the development of agriculture; engineering practices and water management; the movement of people and ideas; and sacred sites, often tied to topography, especially on mountains or hilltop sites.

The chain of continuity in agriculture represents the first theme of the intimate connection between humans and their immediate environment, its evolution over time from the hunter-gatherers to the early farmers, and the representation of a sustainable way of life (Fowler, 2002). Towns or villages may be linked with an agricultural hinterland, indicating the symbiotic relationship between the local dwellers and the land they tilled for sustenance.

Since water has been scarce or limited, innovative technologies, including harvesting and channeling, developed to support agriculture and sustain the communities that settled in these arid or semi-arid regions. This second theme is evident in the relict engineering techniques and topographic interventions aimed at capturing and storing runoff for irrigation purposes, such as those utilized in Nabataean agriculture in the Negev and the agricultural terraces in the Judean Mountains. Nineteenth and twentieth-century irrigation techniques can also be seen in the Sharon region's orange groves, which supported a vital sector of the economy.

Human communications throughout the ages and the movement of people and ideas over land and sea is the third theme of these cultural landscapes, given Israel's geographic location as an international crossroad. Examples include the ancient port cities on the Mediterranean coast and the incense route that traverses the Negev Desert.

The fourth theme relates to dominant topographical features, such as mountains, signifiers of longstanding religious traditions. Many miraculous events occurred on mountain tops, and the reverence towards Mount Tabor, Rosh HaCarmel, and Mount Meron are a few examples of associative landscapes.

Additional narratives include the nineteenth and twentieth-century exemplars of the new planning ideologies and spatial forms that emerged as part of the Zionist movement, such as the cooperative agricultural settlements of the kibbutz and the moshav. Nahalal, for example, was a moshav in the Jezreel Valley with a unique circular plan reflecting its egalitarian social ideology. Other sites include commemorative landscapes designed at the burial sites of leaders of this period, such as Baron Rothschild, whose tomb at Ramat HaNadiv lies in the heart of an elaborate garden, and David Ben Gurion, Israel's first prime minister, who is buried in a landscaped memorial in the Negev desert at Sde Boker.

3.3 Cultural Landscapes: Values and Significance

3.3.1 Level of Significance

After identifying the sites according to landscape categories, the steering committee classified the landscape narratives according to three levels of significance: *universal significance*, landscapes inscribed on the World Heritage List and those worthy of inclusion on Israel's Tentative List; *national significance*, landscapes worthy of conservation in national outline plans; and *district significance*, landscapes worthy of conservation in district outline plans.

The UNESCO terminology was adapted to match the appropriate level of significance. Thus, in this document, the terms "national" and "district" were substituted for "universal": "(a landscape) directly or tangibly associated with events or living traditions, with ideas, or with beliefs, with artistic and literary works of outstanding national significance" (UNESCO, 2021). It was acknowledged that determining the level of significance may be a subjective process that changes over time but provides guidance for the registration of the cultural landscapes within the local planning processes.

3.3.2 Integrity and Authenticity

Two conditions are essential for a landscape to meet the definition of "outstanding value:" integrity and authenticity as indicated in Chapter II.E of the Operational Guidelines. The tests of integrity and authenticity for the inscription process were adopted in Nara, Japan, in 1994. Authenticity can be understood according to attributes including: "form and design; materials and substance; use and function; traditions and techniques; location and setting; and spirit and feeling, and other internal and external factors"[1] (The Nara Document on Authenticity, 1994; OG, 2021). Cultural landscapes must be examined within their local geo-cultural context and based on complete documentation. "Decisions regarding the values of cultural sites and the reliability of information sources may differ among cultures and even within the same culture. Therefore, the determination of values and authenticity cannot be based on fixed criteria; instead, they must be judged within their respective cultural contexts" (Nara, 1994).

"Integrity is a measure of the wholeness and intactness of the natural and/or cultural heritage and its attributes. It addresses three conditions:

1. All the elements leading to the declaration of the site must possess outstanding universal significance.
2. The site must be large enough to ensure a complete representation of the features and processes which convey the property's significance.
3. The site does not suffer from adverse effects of development and/or neglect." (UNESCO, 2021, para 88)

Integrity relates to the attributes that carry the values of a property and refers to their *quantity* concerning a "whole." The question of integrity can be expressed in the following terms: Is there enough left of the property so that we can understand and appreciate its character and meaning?

Authenticity relates to the attributes that carry the values of a property and refers to its *quality* for a given statement of

[1] See Operational Guidelines (UNESCO, 2021), p. 107: "Properties may be understood to meet the conditions of authenticity if their cultural values (as recognized in the nomination criteria proposed) are truthfully and credibly expressed through a variety of attributes." (Paragraph 82).

significance. The question of authenticity may be expressed in the following terms: Is the property really what its given statement of significance claims it to be?

3.3.3 Community Engagement

Who defines Israeli cultural landscapes? Stakeholder coalitions have emerged, bringing together governmental and other national actors with members of local communities to promote initiatives that protect cultural landscapes. Nature conservation has become rooted in civil society in Israel following the establishment of the Society for the Preservation of Nature in 1954—but public awareness of the value of post-antiquity architectural heritage (post 1700 CE), has lagged. The preservation movement gained traction only after many well-regarded historical landmarks were demolished in the rush to build high-rise towers (Amit-Cohen & Amram, 2002; Mualam, 2009). In 1969, the Planning Department of the Ministry of Interior published a survey entitled *Settlement Sites in Israel* by Elisha Efrat and Ehud Gavriely, which surveyed heritage sites from the nineteenth and early twentieth centuries throughout Israel. There had been little awareness of the historical value of such sites before this survey was published. Only archaeological sites and artifacts dated before 1700, the year determined in the British Administration Antiquities Ordinance of 1920, were legally protected. An extensive national project followed, classifying places according to type and period. Three groups of sites were studied: those with historical importance, those typical of an architectural period, and sites related to a particular event.

Efrat and Gavriely argued for the educational role that these sites play: "[t]hese sites are particularly important for the education of our children and the children of the coming generations" (Efrat & Gavriely, 1969, II). They viewed the preservation of the historic town cores as an intergenerational responsibility and local knowledge was utilized to estimate the value and potential rehabilitation options (Efrat & Gavriely, 1969; Turner & Singer, 2015).

Between the late 1960s and the mid-1970s, there was a burst of civil society involvement in heritage conservation. Substantial areas of the Old City of Jaffa and Tiberias were demolished after the establishment of the State in 1948, and preservation attempts in Safed and parts of Jaffa were not wholly successful (Slae et al., 2012). The Council for a Beautiful Israel, established in 1968, was the first NGO to campaign for heritage preservation. In the 1960s and 1970s, architects who had immigrated to Israel from Britain and North America brought new approaches to historic preservation and began to make an impact (Nitzan-Shiftan, 2017; Turner & Singer, 2015). The formation of the Council for the Preservation of Israel Heritage Sites in 1984 by the Society for the Protection of Nature in Israel led to the development of a grassroots movement for protecting post-1700 sites. The credibility of the public awareness campaigns for protecting nature was the main factor in the Knesset decision, linking culture and nature. Subsequently, the Council became an independent NGO in 2008 and was named The Council for Conservation of Heritage Sites in Israel.

References

Amit-Cohen, I., & Amram, G. (2002). Changes in usage in the Old Center of Tel Aviv and attitudes towards conserving built heritage. *Research of the Geography of the Land of Israel, 16*, 289–316.

Efrat, E., & Gavriely, E. (1969). *Settlement sites in Israel.* Ministry of the Interior. (Hebrew).

Fowler, P. (2002). World heritage cultural landscapes, 1992–2002: A review and prospect. In *Cultural landscapes: The challenges of conservation* (p. 16).

IUCN. (2022) *Glossary of definitions.* Retrieved Aug 8, 2022 from https://www.iucn.org/sites/dev/files/iucn_glossary_of_definitions_en_2022.04.pdf.

Mualam, N. (2009). Debating historic preservation in Israel. *PaRDeS: Zeitschrift der Vereinigung für Jüdische Studien eV*, (15), 94–111.

Nara Document on Authenticity (1994). Nara conference on authenticity in relation to the world heritage convention, Nara, Japan. Retrieved Aug 30, 2022, from https://www.icomos.org/charters/nara-e.pdf.

Nir, D. (1970). Geomorphology of Israel. Jerusalem: Akadamon

Nitzan-Shiftan, A. (2017). *Seizing Jerusalem: The architectures of unilateral unification.* U of Minnesota Press.

Slae, B., Kark, R., & Shoval, N. (2012). Post-war reconstruction and conservation of the historic Jewish Quarter in Jerusalem, 1967–1975. *Planning Perspectives, 27*(3), 369–392.

UNESCO (2021). Operational guidelines for the implementation of the world heritage convention. UNESCO: Paris, France. https://whc.unesco.org/en/guidelines/

Turner, M., & Singer, R. (2015). European architectural heritage and conservation in the Levant-Parallels and influences. *Monumenta, 3*, 448–456.

Ziv, B., Saaroni, H., Pargament, R., Harpaz, T., & Alpert, P. (2014). Trends in rainfall regime over Israel, 1975–2010, and their relationship to large-scale variability. *Regional Environmental Change, 14*(5), 1751–1764.

The Israeli Planning Framework

4

Since its establishment, Israel has produced a series of plans that have shaped the country's built environment, infrastructure, and landscapes. The protection of cultural landscapes cannot be understood without a preliminary understanding of the Israeli statutory planning context. This chapter will present the institutional framework that has shaped these plans and discuss how key plans have approached cultural landscapes over time. Master plans at the national and district levels (known as "outline plans"), thematic, or sectoral plans at the national level addressing specific topics and local outline plans have created a statutory framework that has defined the Israeli landscape. This framework has provided the vision and the mechanisms for protecting open space and, more recently, identifying cultural landscapes.

The institutional framework for spatial planning in Israel was established in 1965 with the adoption of the Israel Planning and Building Law. This law defined planning institutions and authorities and established a centralized, hierarchical structure composed of three planning levels or tiers: national, district, and local. Although significant structural reforms have occurred since 2011, land use planning still operates at these three scales. Lower-level local plans must be consistent with all higher-level national plans (Alterman, 2001; Mualam, 2018; Feitelson et al., 2021). Various aspects of cultural landscapes have been addressed at each of the three scales or levels.

4.1 National Outline Plans and Sectoral Plans

In 2005, the National Planning and Building Board approved a new long-term national outline plan, the "Comprehensive National Outline Plan for Construction, Development and Conservation 35," (known as *TAMA 35* in Hebrew, or NOP 35) that had far-reaching effects on Israeli planning. It created a new planning language based on the concept of "textures," or "fabrics," based on five texture typologies: urban texture, rural texture, mixed preserved texture, national preserved texture, and coastal texture areas. Statutory instructions were developed for each texture to regulate land uses and provide development and conservation guidelines. The plan created a classification system for all landscape types in the country as "landscape ensembles." Notably, these included cultural landscapes. Thirty "landscape ensemble" polygons were designated, "mostly open spaces that are highly sensitive and combine landscape values and human interventions."[1]The first amendment to NOP 35 adjusted the boundaries of the ensembles, partially in response to the publication of the Hebrew edition of this book in 2010.[2]

Sectoral national plans on specific themes, such as roads, railroads, tourism, heritage sites, and nature reserves, provide detailed land use regulations and focus on the significance of natural and cultural assets. These plans addressed the preservation of landscape features and systems to various degrees. The lack of clear and consistent definitions has resulted in overly open-ended limitations on development. However, specific plans, such as the 1970s plan for parks and nature reserves, have been critically important to preserving remaining regional open spaces such as the Carmel Mountain range. Other plans were critical to protecting forests and coastlines (Alterman, 2001). Recently, some 300 individual sectoral national outline plans were combined into a single plan known as National Outline Plan 1, approved in 2020.

[1] These were defined by landscape architect Shlomo Aronson, a NOP 35 planning team member.

[2] See Document 2 of the 2016 Amendment to NOP 3 https://www.gov.il/blobFolder/generalpage/tama_35_docs/he/National_planning_institutions_tama_35_2.pdf.

A. Sar Shalom et al., *Cultural Landscapes of Israel*,
https://doi.org/10.1007/978-3-031-33685-0_4

4.2 District Outline Plans

District Outline Plans are prepared by the District Planning Commissions for each of Israel's six administrative districts and are subject to the approval of the National Planning and Building Council. They are meant to interpret the National Outline Plan at the district level by addressing district-level land uses such as new or renewed neighborhoods, industrial areas, and open spaces. Thus, they can provide a holistic, regional perspective that mediates between national and local scales and is the most critical scale for preserving open space systems and ensuring their connectivity (Feitelson et al., 2010).

Many District Outline Plans addressed landscape contexts, including natural features like streams. The cultural landscape components of the district plans, just as in the national plans, are generally referenced only in the sections dealing with the environment or conservation. For example, the partial district plan of the Central District (District Outline Plan 3/10) for the Yarkon River (*Nachal HaYarkon*) does not mention the cultural landscape, despite the presence of physical relicts and historical sources from various periods that indicate the long presence of human activity near the river. The only references to the cultural landscape in this plan relate to agricultural uses.

Tangible references to cultural landscapes can be found in the Partial District Outline Plan for the Nadiv Valley (Bika'at HaNadiv) (2007), created after the enactment of National Outline Plan 35. This was the first in a series of plans that approached landscape management as an issue in its own right. The Nadiv Valley plan defines the "cultural landscape ensemble" as a functional landscape with distinct ecological values of varying sensitivity. The plan designated an ecological corridor to restrict development and promoted the preservation of traditional agriculture, touristic and recreational sites, and scenic roads. The plan was ultimately defeated in 2018 in response to the opposition of local farmers. They considered the plan's limitations of industrial agricultural elements such as greenhouses and other structures overly restrictive.[3]

In 2018, the Planning Administration regrouped some 250 district plans and amendments and simplified the multiple landscape definitions using only two categories, agricultural and natural, without directly referencing the cultural landscape.[4]

[3] https://www.calcalist.co.il/articles/0,7340,L-3843635,00.html.

[4] https://www.gov.il/BlobFolder/generalpage/tammim/he/tmmim.pdf.

4.3 Other Planning Frameworks

In addition to the Planning and Building Law, Israel has a range of laws granting protection to heritage sites within clearly delineated boundaries, such as: the Protection of Holy Places Law (1967), the Antiquities Law (1978), National Parks, Nature Reserves, National Sites and Memorial Sites Law (1998). Over one hundred sites have been defined as "sacred sites," while a few others have been declared "national commemoration sites." Local authorities have defined hundreds of local "historic sites" beyond the definition of antiquities and thousands of additional sites that have not yet been officially declared.

The viability of cultural landscapes depends on a broad coalition of support. Many government ministries play a critical role in protecting and promoting cultural landscapes, including the Planning Administration and the Ministry of the Environment. The Israel Nature and Parks Authority (INPA) was tasked with enforcing the National Parks, Nature Reserves, National Sites and Memorial Sites Law (1998) and managing the National Parks and Nature Reserves. If the area in question includes a holy place or a historical site, the Ministry of Religious Affairs or the Ministry of Education and Culture has jurisdiction over the site and is charged with ensuring its protection.

The Ministry of Agriculture is involved if the area in question is located in a nature reserve; the Ministry of Defense is involved if the area, or a section of it, is declared a military zone. The Israel Land Authority, the government agency charged with managing state land (which comprises some 93% of the country's total area), is another significant stakeholder.

Although current legislation emphasizes built heritage, non-governmental organizations (NGOs) such as the Council for Conservation of Heritage Sites in Israel and the Society for the Protection of Nature in Israel (SPNI) have played essential roles in generating public awareness of Israeli natural and cultural heritage and the broader meaning of cultural landscapes. Landscape archaeology, for example, is a concept that has yet to be formally accepted. The listing of the Horns of Hattin is a first step in bringing forward archaeologically rich landscapes toward mainstreaming as cultural landscapes. Environmental NGOs have significantly influenced policy and provided public outreach and education on environmental issues. They have also spurred civil society efforts on a range of initiatives. These organizations have served as watchdogs on government initiatives, and on many occasions, they have been vocal opponents of plans perceived as threats to the environment. The battle for Gazelle Valley Park is one such example. Community groups, together with the Society for the Protection of Nature (SPNI), successfully opposed the development of a new neighborhood on an ecologically sensitive site in Jerusalem. After a lengthy court battle, an urban wildlife park was implemented to protect the existing herd of mountain gazelles. Other initiatives include coastal protection, fishing reduction initiatives, stream restoration, air pollution reduction, and the protection of open space and wildlife in numerous locations.

And yet, these efforts are insufficient for preserving cultural resources. Local interventions to safeguard heritage sites cannot ensure effective stewardship unless the surrounding context is considered. The approach to cultural resource conservation in Israel today is reminiscent of the approach to nature conservation several decades ago when nature reserves were planned as separate, disconnected entities, defined by a statutory "wall" that did not address the surrounding context. Lessons learned from the field of biodiversity have significantly expanded planners' practical "toolbox." Research advances have resulted in greater conceptual sophistication, highlighting the importance of elements such as wetlands or ecological corridors for maintaining biodiversity, which have, in turn, helped define more effective planning tools. The nature reserve, for example, is no longer the only conservation option; other tools have emerged, such as biosphere reserves or urban natural areas, among others. These concepts have yet to be fully incorporated into Israeli practice. New approaches must be developed to expand the current toolkit to define and protect Israel's diverse and often contested cultural landscapes.

4.4 Defining the Sites: Mapping Methodologies

In order to translate selected narratives from a qualitative document described in words to quantitative data expressed visually on a map, a process was developed to define site boundaries, or polygons, for each cultural landscape. With the importance of the wider setting, a "buffer zone" was added around the cultural landscape polygon as an added layer of protection. The definition of the polygon boundaries and buffer zones was based on the geocultural realities and did not take into account the development plans proposed for each area. Nonetheless, we believe it was significant to include these areas as cultural landscapes even in such cases, to minimize the impact of change. For example, a plan for a major highway and rail intersection in the Tel-Kashish area that crossed an area classified as an associative landscape of universal significance (Elijah and the Prophets of Ba'al) was modified as a result of the designation. The database did not include cultural landscapes whose heritage values were already being destroyed by development. For example, Tel-Yarmut is surrounded by existing and planned development initiated by the town of Beit Shemesh. It was, therefore, not included in the cultural landscape entry of "A row of *tels* in the Judean Hills (Shfela)" as it was no longer possible to differentiate the *tel* from its surroundings. Only *tels* that had a distinct presence in the landscape were included in the database.

4.4.1 Cultural Landscape Polygons

The cultural landscape polygons represent the geographic boundaries of the "narratives" that constitute the cultural and natural heritage expressed in the landscape. The physical delineation of the narrative varied according to its category: designed landscape, organically evolved landscape, or associative landscape. The delineation process was most straightforward in a designed landscape, where the boundary corresponded to the limit of the intervention. It may be marked by a fence or another type of clear physical boundary, as seen in the Tomb of the Baron de Rothschild (Ramat HaNadiv). These landscapes were generally small, and even if the landscape design was no longer clearly defined, the relicts of the designed area are still often clearly identifiable within the landscape context, such as at the Gilgal Refaim site in the Golan Heights.

The delineation of organic landscapes includes tracing relicts in the field and undertaking cartographic work to establish thematic groupings, such as olive groves. In cases where the cultural heritage is evident only through historic relicts that represent traditions and lifestyles that are no longer practiced, archaeological elements are identified, for example, in the relicts of ancient agriculture in the Negev (see 5.1.1.2).

The most complex delineation process concerned translating a more abstract associative landscape into concrete geographic terms. This work was done through close observation and in consultation with experts in the field who had experience conveying the narrative to different audiences. This process involved identifying the vantage points and specific landscape areas that could bring the story to life. After this subjective information was placed on a geometrically corrected aerial photograph, it was interpolated to create the relevant points and areas in a single polygon.

Some cultural landscapes were represented in different sites as serial designations with physical relicts occurring in multiple locations. The Operational Guidelines on serial properties were adopted *mutatis mutandis* in the organic landscape of the terraced agriculture in the Judean Hills, for example (see 5.1.3.10), and the designed landscape of the avenues of *Washingtonia* palms (see 5.2.1.3).[5] This latter serial designation was created following a preliminary selection process. Only three examples were included to avoid overshadowing the clearest and most complete examples that met the conditions of authenticity and integrity.

4.4.2 Cultural Anchors

In many defined cultural landscape polygons, specific elements anchor the narrative within the landscape. These cultural anchors are the points that demonstrate the landscape's cultural heritage most clearly or highlight criteria that enable the narrative to be included in this database. They may be focal points that provide the raison d'être for the site as a sacred place; or centers of social activity, such as the historic dining room structure at the center of the Workers' Camp in Sodom, Dead Sea (see 5.2.2.7). They may also provide an overlook of the wider landscape, enabling the viewer to gain a broader perspective on the cultural heritage of the place.

4.4.3 Buffer Zones

Based on the Operational Guidelines of the World Heritage Convention, buffer zones are areas that provide an added layer of protection, including the immediate setting, important views, or other areas or attributes.[6] Often, the buffer zone area is more expansive than the cultural landscape polygon itself and is a tool used to sustain and protect the context and setting surrounding the cultural landscape. It is established by determining the most significant values present in the cultural landscape and deciding which steps are needed to maintain their integrity and authenticity.

In some cases, the buffer zones occur in the viewshed of the cultural landscape. This is particularly important where temporal detachment is necessary to experience the narrative. The experience can be intensified using spatial means, as in the Incense Route—Desert Cities of the Negev and the Beit Netofa Valley. The buffer zone was set along the ridgeline that frames the cultural landscape in these examples. In other cases, only the views of the landscape directly related to the narrative were addressed. For example, while there are expansive views of the Carmel, Samaria and the Lower Galilee from the top of Tel Megiddo, the Armageddon narrative relates only to the plain to the north of the *tel*.

The buffer zone can also be used as a connecting corridor between the cultural landscape and adjacent open spaces to prevent the landscape from becoming completely isolated by encroaching development. An example can be seen at Mount Tabor, where Jesus chose to seclude himself and where the small open spaces between the villages were defined as buffer zones to prevent the spread of development around the mountain.

The essence of the associative cultural landscape is to convey more abstract cultural values associated with a region by setting it off from its surroundings. Nevertheless, it is very challenging to define the exact boundaries of such an area since it is not derived from specific physical elements on the site. For example, this was an issue in the Gilboa region, where King Saul's battle against the Philistines is believed to have occurred some three thousand years ago. As

[5] See the Operational Guidelines (UNESCO, 2021, Paragraph 137) for the definition of a serial property: "Nominated serial property includes two or more component parts related by clearly defined links: (a) Component parts should reflect cultural, social or functional links over time that provide, where relevant, landscape, ecological, evolutionary or habitat connectivity; (b) Each component part should contribute to the Outstanding Universal Value of the nominated property as a whole in a substantial, scientific, readily defined and discernible way, and may include, inter alia, intangible attributes. The resulting Outstanding Universal Value should be easily understood and communicated; (c) Consistently, and in order to avoid an excessive fragmentation of component parts, the process of nomination of the property, including the selection of the component parts, should take fully into account the overall manageability and coherence of the nominated property" (see Paragraph 114).

[6] See the Operational Guidelines (UNESCO, 2021 Paragraph 104) for a definition of buffer zones: "For the purposes of effective protection of the nominated property, a buffer zone is an area surrounding the nominated property which has complementary legal and/or customary restrictions placed on its use and development to give an added layer of protection to the property. This should include the immediate setting of the nominated property, important views and other areas or attributes that are functionally important as a support to the property and its protection. The area constituting the buffer zone should be determined in each case through appropriate mechanisms. Details on the size, characteristics and authorized uses of a buffer zone, as well as a map indicating the precise boundaries of the property and its buffer zone, should be provided in the nomination."

stated in the Operational Guidelines, "Attributes such as spirit and feeling do not lend themselves easily to practical applications of the conditions of authenticity, but are important indicators of character and sense of place, for example, in communities maintaining tradition and cultural continuity" (UNESCO, 2021, p. 31).

The cultural landscapes identified in this database perform as buffer zones between developed areas and the virtual locations in which the narratives are represented.

References

Alterman, R. (2001). National-level planning in Israel: Walking the tightrope between government control and privatisation. In *National-level planning in democratic countries: An international comparison of city and regional policymaking* (pp. 257–290).

District Outline Plan 3/10 for the Yarkon River (2008).

Feitelson, E., Felsenstein, D., & Razin, E. (2010) Evaluating the district plan in Israel. A report to the planning administration. (Hebrew).

Feitelson, E., Horowitz-Harel, A., Levin, N., Mintz, Z., Steenekamp, G., & Zaban, S. (2021). Haste makes waste: On the implications of rapid planning in Israel. *Land Use Policy, 103*, 105312.

Mualam, N. (2018). Playing with Supertankers: Centralization in land use planning in Israel—A national experiment underway. *Land Use Policy, 75*, 269–283.

National Outline Plan 3: Roads, 1991. Revised 2005.

National Outline Plan 8: National Parks and Nature Reserves (1981) . Revised as Tama 1, 2020.

National Outline Plan 13: Mediterranean Coastlines. Revised 2021.

National Outline Plan 22: Forest and Afforestation (1999). Revised as Tama 1, 2020.

National Outline Plan 23: Railroads (1986).

National Outline Plan 35. (2005). Revised 2016.

The National Outline Plan 12: First Amendment for Tourism and Recreation. (1999). Revised 2020.

UNESCO (2021). Operational guidelines for the implementation of the world heritage convention, UNESCO World Heritage Centre, Paris. https://whc.unesco.org/en/guidelines/.

5 Case Studies

This chapter presents a list of fifty cultural landscapes in Israel that were identified, researched, and mapped by the Israel Nature and Parks Authority. They are organized according to the following classification: properties inscribed on the World Heritage List, properties noted on the World Heritage Tentative List, Cultural Landscapes with National Levels of Significance and Cultural Landscapes with a District Level of Significance. Properties inscribed on the World Heritage List are subdivided into cultural landscapes and other properties. Cultural landscapes with a National Level of Significance are further subdivided according to the categories of the World Heritage Convention: (i) designed landscapes; (ii) organically evolved landscapes (relict and continuing); and (iii) associative landscapes. (See Appendix C).

Each cultural landscape includes a statement of significance, a narrative text describing the cultural values worthy of safeguarding and its geocultural extent. This is established by a polygon delineated in yellow on the aerial photo. Serial cultural landscapes are represented by multiple polygons. The scale varies according to the size of the site. The buffer zones, representing the landscape and wider setting, were defined around most polygons and delineated in blue. The cultural anchors are marked by a yellow square in which their names are listed.

A national map of the proposed cultural landscapes was drawn to a scale of 1:450,000. Each category is represented by a different color: designed landscapes are shown in blue, continuing organic landscapes are purple, relicts of organically evolved landscapes are green, and associative landscapes appear in brown (Fig. 5.1).

This preliminary list is intended to provide a basis for integrating cultural heritage into the statutory planning processes.

As of 2022, Israel has nine properties inscribed on the World Heritage List. Four were included in the national cultural landscape database compiled by the Israel Nature and Parks Authority. However, only one of these four, the Incense Route—Desert Cities in the Negev, was listed specifically as a cultural landscape. The other three properties were nominated and listed based on different criteria though they are also considered cultural landscapes at the national level. They include the Bahá'i Holy Places (Haifa and the western Galilee); the Sites of Human Evolution at Mount Carmel (The Nahal Me'arot/Wadi el-Mughara Caves); and the Caves of Maresha and Bet-Guvrin (Judean Lowlands).

Tel Megiddo is a part of an additional World Heritage site included in the national Cultural Landscapes Inventory. It is a component of the biblical *tel* serial nomination. It consists of a futuristic cultural landscape looking from the *tel* (archaeological mound) towards Armageddon, which according to the biblical prophets, is the battle to end all times.

5.1 Cultural Landscapes with a Universal Level of Significance

5.1.1 Inscribed on the World Heritage List as Cultural Landscape

5.1.1.1 The Incense Route: A Relict Organically Evolved Landscape (Figs. 5.2, 5.3 and 5.4)

Narrative

The Incense Route originates in Oman and Yemen, and its total length extends over 2400 kilometers. The route passes through Saudi Arabia and Jordan, traverses the Negev and ends at the port of Antidon (Gaza). The route and extensions bustled with life between the third century BCE and the fourth century CE. The route was primarily used to transfer frankincense and myrrh, though additional commodities

The original version of the chapter has been revised: Belated corrections for chapter footnotes have been updated in page 46 and 78. A correction to this chapter can be found at https://doi.org/10.1007/978-3-031-33685-0_8

A. Sar Shalom et al., *Cultural Landscapes of Israel*,
https://doi.org/10.1007/978-3-031-33685-0_5

Fig. 5.1 National map—cultural landscapes

such as spices, textiles and metals were also transported along the route. The destination was Greece and the Roman Empire, and Antidon and Rhinocorura (El Arish) ports served as the sea link to Europe.

The Roman historian Pliny the Elder described the route during the first century CE: "After it is collected, frankincense is conveyed to Sabota on camels, one of the gates of the city being opened for its admission; the kings have made it a capital offense for camels so laden to turn aside from the high road. At Sabota, a tithe estimated by measure and not by weight is taken by the priests for the god they call Sabis, and the incense is not allowed to be put on the market until this has been done; this tithe is drawn on to defray what is a public expenditure, for actually on a fixed number of days the god graciously entertains guests at a banquet. It can only be exported through the country of the Granite, and accordingly, a tax is paid on it to the king of that people, as well. Their capital is Thomna, which is 1487½ miles distant from the town of Gaza in Judea on the Mediterranean coast; the journey is divided into 65 stages with halts for camels. Fixed portions of the frankincense are also given to the priests and the king's secretaries. Still, besides these, the guards and their attendants and the gate-keepers and servants also have their pickings: indeed, all along the route, they keep on paying, at one place for water, at another for fodder, or the charges for lodging at the halts, and the various octrois; so that expenses mount up to 688 denarii per camel before the Mediterranean coast is reached; and then again payment is made to the customs officers of our empire. Consequently, the price of the best frankincense is six, of the second best five, and the third best three denarii a pound" (*Historia Naturalis*, Pliny the Elder, Book 12: XXXII).

The Landscape

The section of the Incense Route that traverses Israel is 180 kilometers long, extending from the Jordanian border to Gaza. The section from Moa in the Arava to Avdat in the Negev Highlands was declared a UNESCO World Heritage Cultural Landscape in 2005. This section contains remains, including the city of Avdat, strongholds, a road with paving stones, milestones, caravanserai and sophisticated agricultural systems. On the section of the road that extends north of Avdat towards Gaza, little has been found to mark the route. Indeed, only two points marked with milestones were found north of Avdat. The main point of interest in this portion of the route is the city of Halutza, which was also included as part of the UNESCO declaration.

The remains along the declared route are an excellent example of cultural landscapes. Against the background of the most impressive landscapes of the Negev desert, with its

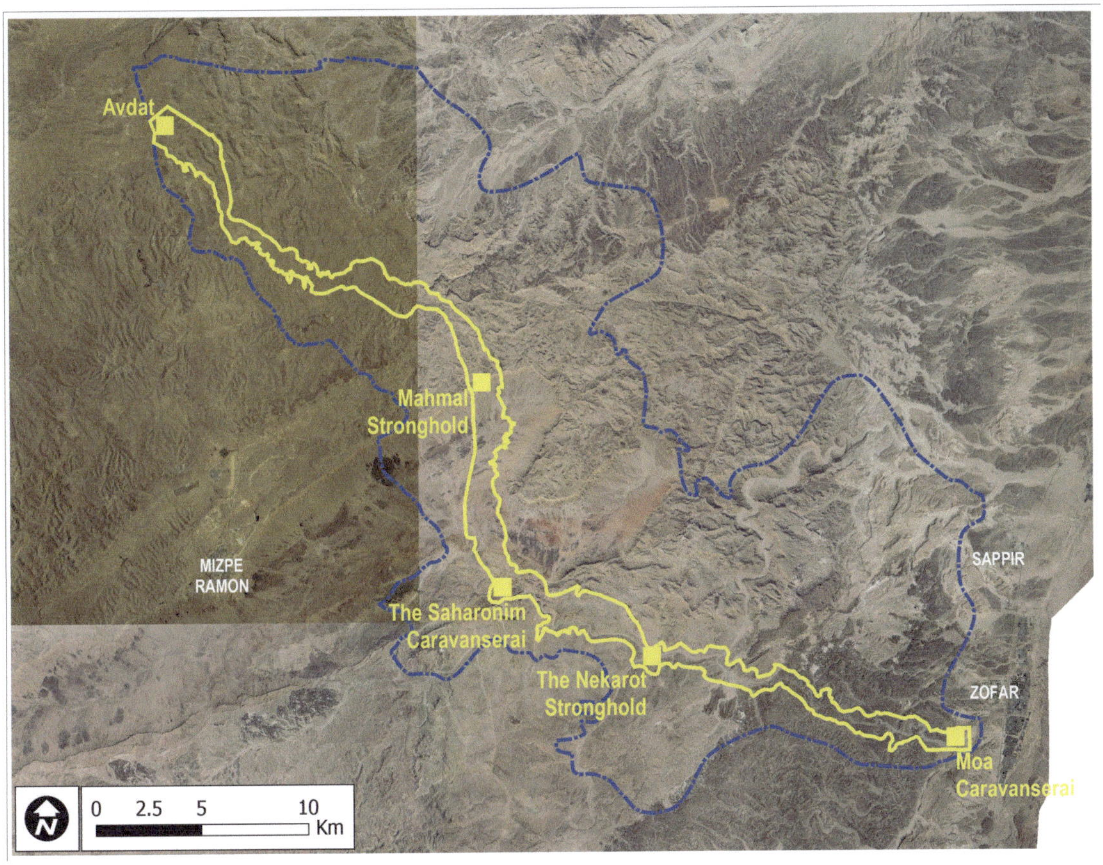

Fig. 5.2 The Incense Route—polygon

mountains, riverbeds and desert vegetation, it is possible to see fascinating remains arranged systematically along the route. The Moa caravanserai was the "60th station" along the way and served as a place to stay the night.

Near the large caravanserai is a spring that provides a constant water source, ensuring the ability to host large caravans. The stronghold, Kazra, is located adjacent to a cistern eleven kilometers away, and ten kilometers further is the Nekarot stronghold (Metsad Nekarot) that guards the Nekarot Ascent (Ma'ale Nekarot). This section passes over the Nekarot stream and continues towards the Ramon Crater (Makhtesh Ramon). The Nekarot stronghold similarly has a cistern in its vicinity.

In these sections, it is possible to see numerous remnants of cairns or other arrangements of stones along the way. Some of them are landmarks, while others represent worship. Traveling another eleven kilometers along the route, one arrives at Khan Saharonim, the next place to spend the night, and the "61st Station." The Saharonim Spring provides a constant source of water. Further on, the route traverses the Ramon Crater from south to north, with stones lining the road and milestones scattered along this section. The crater (*makhtesh*) offers travelers scenic views that are colorful, sandy and rocky with a dramatic backdrop of the cliffs; after an additional eleven kilometers, one reaches the Mahmal stronghold.

The route proceeds north and reaches the expansive Nafcha Heights (Ramat Nafcha), a well-preserved section of the route. This section is fifteen kilometers long. Approximately five meters wide, the road is marked with prominent stones on both sides and, at some points, with milestones. As we approach the city of Avdat, the terraces are more concentrated, indicating the ancient agricultural hinterland surrounding the city. The combined length of the route is 65 kilometers: two days of walking. Every caravanserai has a permanent water source, while the strongholds between the caravanserais have only cisterns that collect runoff. While the first section of the route is an arid desert, the second section (from the Mahmal stronghold and northward) has a climate that supports agriculture, as evidenced by the remains of ancient agricultural systems.

Fig. 5.3 The remains of a cistern on the Incense Route

Fig. 5.4 The remains of a stronghold along the Incense Route

Defining Polygons

The cultural landscape polygon defines the area designated as a World Heritage Site. Between the Moa caravanserai in the Arava and the city of Avdat, the remains of the road are clearly defined. Today travelers can still experience the landscape crossed by traders for over 700 years. The buffer zone includes the ridgeline that commands the road and frames the area. Any change to the buffer zone would affect the experience of crossing the desert as it might have seemed to the traders as they walked in the fourth century CE.

Cultural Anchors

Moa Caravanserai—A stronghold, bathhouse and caravanserai used for lodging. The stronghold sits on a hill, dominating the small valley where the caravans would concentrate. A bubbling spring filled an ancient pool that had a bathhouse built alongside it. Inside the stronghold, an olive press for producing olive oil was discovered.

The Nekarot Stronghold—on the north bank of the Nekarot stream, lay the remains of Nabataean and Roman-era buildings once used to secure the road. Three hundred meters south of the stronghold is a cistern used by the inhabitants of the stronghold. This cistern is one of the best-preserved examples of water reservoirs that have survived in the Negev from the Nabataean period.

The Saharonim Caravanserai—The "61st Station" on the Incense Route is a square building with a single-entry gate that leads into a large courtyard surrounded by rooms. Some rooms were used for sleeping; others were workshops for artisans, washing tubs and an oven.

Mahmal Stronghold—a watchtower with dimensions of 6.5 by 7 meters, consists of two rooms. It was situated on the top of the Mahmal Ascent overlooking the Ramon Crater. Approximately 500 meters north was a water reservoir measuring 4.8 by 8 meters that could hold 150 m^3 of water.

Avdat—The "62nd Station," initially founded as a way station on the Incense Route, blossomed into a city containing Nabataean and Roman temples that later became churches, a residential district, camel pens, workshops, a cistern, and a bathhouse.

5.1.1.2 Ancient Agriculture in the Negev: A Relict Organically Evolved Landscape (Figs. 5.5 and 5.6)

Narrative

For a thousand years, a complex system of settlement existed in the Negev that has left a lasting imprint on the landscape. Settlement in the desert began with the arrival of the Nabataeans to the Negev in the third century BCE. The Nabataeans were nomadic traders who originated from the Arabian Peninsula. They established their capital in the city of Petra. According to Hieronymus of Cardia, the Nabataeans avoided planting trees, sowing fields, and constructing permanent structures. However, they laid the foundations for way stations and commerce throughout the Negev Desert. With the annexation of the Nabataean kingdom by the Roman Empire in 106 CE, a process of transformation began, and the way stations established by the Nabataean expanded into permanent settlements. These cities reached their zenith during the Byzantine period, between the fifth and seventh centuries CE, and were abandoned after the Muslim conquest in 636 CE.

A number of these stations became permanent settlements. Mamsheet, Halutza, Shivta, Nitzana, and Avdat, are usually called the ancient cities of the Negev. These six Nabataean cities have archaeological remains, indicating the profound change Negev inhabitants underwent. Initially nomads, the Nabataeans evolved into permanent town dwellers, farmers and builders who worshiped in Christian churches. The agriculture that developed in these cities was impressive in its scope. It relied on the collection of rainwater and its diversion to irrigate crops. These impressive waterworks included channels, reservoirs and dams, which enabled ancient farmers to control the water quantities allocated to every plot. Surprisingly, the ratio between the surface area irrigated and the area from which the runoff was collected stood at an average of 1:20 and significantly increased the amount of water available for cultivated plants.

Each city has a different geographic setting and distinct attributes; some cities are on the plain, while others are on the hilltop, overlooking their surroundings. However, they share several characteristics: they are located near riverbeds as a water source for agriculture and on the trade route, alluding to their origins as Nabataean trading stations.

The Landscape

Cities that developed agriculture using the ancient water collection systems were located in the western part of the Central Negev, from the Plain of the Winds (Mishor Ha-Ruach) in the south to Mamsheet in the north and Nitzana in the west.

It is possible to identify the remains of the systems used to collect rainwater over an extensive area around these ancient cities. Though the agricultural systems have been abandoned for millennia, some are still partially functioning. Bedouins sporadically use a few as they cultivate their fields. In other areas, this system supports a high concentration of wild plants or a dense cluster of flora that exceeds what is naturally found in the local arid climate.

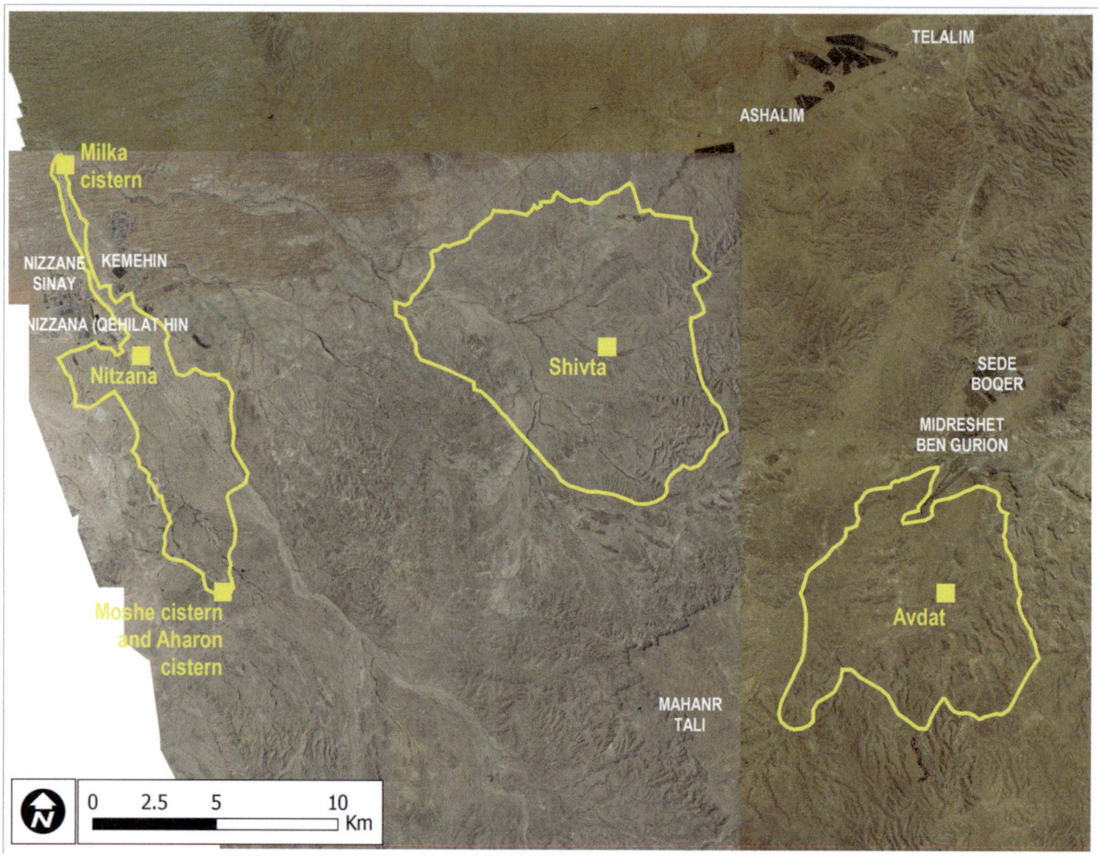

Fig. 5.5 Ancient agriculture in the Negev—polygon

Fig. 5.6 Ancient agriculture in the Negev near Tel Avdat.

The heart of these agricultural systems are terraces constructed in the stream valleys used to cultivate vineyards and orchards by collecting runoff. The ancient water cisterns can often be identified by a large tree that has grown where the water still flows through this partially preserved system. Large dams constructed to hold back floods still exist along the drainage ditches. Near the cities, one can find wine presses, olive presses, granaries and other facilities related to processing agricultural products. These remains testify to the early crops cultivated in these fields: grapes for wine, olive oil and grains. The Even-Ari farm near the city of Avdat and the Shivta Farm were set up to restore ancient agricultural methods and an attempt was made to utilize these methods to adapt modern agriculture to extreme desert conditions. These two farms, located in the heart of an ancient agricultural region, partially use the original system. Thus, they are a relict of the organic landscape while continuing their role as an organic agricultural landscape to this day.

Cultural Anchors

Moshe cistern and Aharon cistern, Nitzana, Milka cistern, Shivta.

5.1.2 Inscribed on the World Heritage List

5.1.2.1 The Bahá'í Gardens: A Designed Landscape as Part of the Bahá'i Holy Places in Haifa and the Western Galilee (Figs. 5.7, 5.8, 5.9, 5.10 and 5.11)

Narrative

The Bahá'í religion was established in 1844 in the city of Shiraz in Persia by the Báb ("the messenger"), Siyyid `Alí-Muhammad. The Báb decreed that his role was to pave the way for the arrival of the great prophet who would open a new chapter for humanity, ushering in an era of peace, prosperity and unity. Thousands of Shiite Muslims became followers of the Báb. The Báb himself was imprisoned and eventually executed in 1850. The arrival of the heralded prophet appeared shortly after that in the form of Mírzá Husayn Alí Núrí, called the Bahá'u'lláh (the glory of God), and his followers are named after him—Bahá'í. The Persian authorities feared this new movement and exiled the Bahá'u'lláh to a prison in Acre.

The Bahá'u'lláh was buried in Acre, his son, Abdu'l-Bahá, followed his path, and in 1908 he named the city of Haifa the holiest Bahá'í city. The slope of Mount Carmel, facing the tomb of the Bahá'u'lláh in Acre, subsequently became the religious and administrative center of the Bahá'í community. The temple of the Báb ("the Golden Dome"), the symbol of Haifa, was built in 1953. The Báb's bones were moved from Acre to the temple, and in 2001 the surrounding grounds were transformed into magnificent hanging gardens sculpted into the side of the mountain.

The Landscape

The Bahá'í gardens are "hanging" gardens on Mount Carmel's steep slope extending over a kilometer. The Louis Promenade overlooks the gardens, the picturesque German Colony and Haifa Bay. Together these elements create a rare landscape ensemble. The garden includes eighteen terraced gardens along the slope. Nine gardens lie below the Golden Temple, and nine lie above it (the number nine in the Bahá'í faith represents the major world religions). Every terrace is constructed as a flat plane paved with stones. On either side, water gathers in pools and rises up in fountains. The water continues its calming flow down the mountain to the lowest terrace. The water is pumped back up to the top to trickle down again. The gardens provide a fitting backdrop for the grand temple.

Palm trees on each terrace create a sense of continuity for visitors descending through the gardens. Each terrace contains decorative gardens, fountains, embellished railings and statues encircled by verdant grassy areas, designed to look like nine concentric waves emanating from the temple. At the edge of each terrace, the pine trees and local forest have been preserved and blend with the garden's foliage, serving as a nesting place and habitat for birds and animals. Thus, the gardens have an additional role as a small nature reserve in the heart of industrialized Haifa. The gardens are carefully maintained. The garden designer, noted architect Fariborz Sahba, produced a work of art.

Al Bahaja: The Bahaja garden is situated beside the Acre-Nahariya Road, slightly south of Bustan Hagalil. Ornate iron gates mark the northern entrance to the garden. These gates are called the Kolinus Gates after Emelia Kolinus, the donor. Beyond the gates lies a charming garden in the finest Baha'i style. The flowerbeds are well-tended and linked by paths paved with brown clay gravel. The garden contains

Fig. 5.7 The Bahá'í Gardens—Shrine of the Báb—polygon

sculpted peacocks and eagles. Several impressive araucaria trees are also prominent in the garden. At the beginning of the twentieth century, they were praised by Aharon Aaronsohn, the noted Jewish agronomist and botanist.

A manor house lies in the heart of the garden, built by a wealthy Acre merchant in 1870. The Bahá'u'lláh was allowed to live in a home near Acre while still officially a prisoner of the city. He lived in the house for seven years until his death. The Bahá'u'lláh was buried close to the house, and an elaborate mausoleum was constructed on his gravesite.

Defining Polygons

The polygon of the Bahá'í gardens in Haifa is the area enclosed by the wall surrounding the garden; an additional one-hundred-meter buffer was designated to ensure that surrounding developments would not overshadow the fine qualities of the garden. The polygon for the Baha'i Gardens near Bustan Hagalil is similarly defined as the area surrounded by the garden wall. All the open agricultural land adjacent to the garden has been designated as a buffer zone since one of the garden's primary qualities is its location within an agricultural setting.

Fig. 5.8 The Bahá'í Gardens—Shrine of the Báb

Cultural Anchors

Al Bahaja (the Bahá'u'lláh's house in Acre), Shrine of the Báb (The Golden Dome in Haifa).

5.1.2.2 Megiddo and Armageddon: An Associative Landscape as Part of the Biblical *Tels*—Megiddo, Hatzor, Beersheba (Figs. 5.12, 5.13 and 5.14)

Narrative

Tel Megiddo is the site of one of Israel's most important historical cities. The city on the *tel* commanded a view over one of the critical points of the "coastal road," which originated in the Kini stream and extended to the Jezreel Valley. Twenty-five strata of settlements were discovered at the *tel*, ranging from the Pre-Pottery Neolithic period and culminating in the Persian period. Temples, brick walls and ritual altars were revealed from the earlier periods in a palace from the Canaanite period, a rich collection of ivories with boxes, figurines and plaques depicting figures of gods and animals.

Megiddo was first mentioned in Egyptian inscriptions describing the campaigns of Thutmose III in 1468 BCE when the Canaanite city rebelled against Egypt, finally succumbing after seven months of siege. Megiddo was also an important city during the Israelite period. According to

Fig. 5.9 The Bahá'í Gardens—Shrine of the Báb

the Bible, King Solomon fortified Megiddo along with Hatzor and Gezer (I Kings 9: 15). The city reached its zenith under the rule of King Ahab. During the site's excavations, a large-scale water supply system was uncovered, with an underground tunnel extending eighty meters between the city and a hidden spring outside the walls. This system gave the inhabitants of Megiddo access to water even during a siege.

The New Testament describes Armageddon as the site of the battle of Gog and Magog that will take place at the End of Days. After this battle, peace will come to earth (John 17: 21–14). The source of the name Armageddon is likely Mount Megiddo, and the Christian tradition identifies the end-of-days battlefield at the foot of Mount (*Tel*) Megiddo. In 1918, British General Allenby won a decisive battle at the site against the Ottomans at the conclusion of World War I. He was given the title "Lord Allenby of Armageddon" after the battle.

With the declining importance of the small walled town, the village Othnai was built in the first century CE on the hill south of the Kini stream and Tel Megiddo. It is mentioned in Israelite sources from the first to the fourth century CE. The strategic importance of the source of the Kini stream for the Jezreel Valley was recognized by the Sixth Roman Legion, who established their headquarters on a field northwest of the hill where the Jewish settlement was located, and the site was named Legio. The Roman military presence at the site continued for two centuries, during which the rural settlement absorbed a foreign population, including the Samaritans. The settlement grew until it became a town in the late third century CE when it was renamed Maximianopolis (City of Maximian). A large building was recently uncovered, featuring a mosaic floor in a Christian prayer hall from the third century CE. The mosaic inscription mentions a Roman army officer who donated funds to construct the mosaic when Christians were persecuted in the Roman Empire before Christianity had become the state religion of the Byzantine Empire.

The Landscape

Tel Megiddo, located on the border of the Jezreel Valley and Ramot Menashe, rises only 60 meters above its surroundings. Yet the *tel* provides spectacular views to the north, east and west: across the cultivated fields of the Jezreel Valley, the hills of Nazareth and the mountains of Moreh, Tabor and Gilboa.

The landscape is not only beautiful and diverse, it is also saturated with biblical memory. It is the site of dramatic events described in the Bible, such as the battle between Saul and the Philistines and Barak's battle against Sisera. These are also the landscapes that Christian scriptures designate as the arena for the battle to end all battles, after which peace will come to the land.

Did the Roman officer who contributed to creating the mosaic in the Christian chapel in Maximianopolis share these associations? We will never know, but it is clear that hundreds of millions of Christians worldwide today have strong ties that bind them to these landscapes.

Fig. 5.10 The Bahá'í Gardens—Al Bahaja—polygon

Defining Polygons

The archaeological site of Tel Megiddo is a protected national park that has been recognized as a UNESCO World Heritage site. The cultural landscape polygon includes the *tel*, its immediate surroundings, and archaeological remains from post-biblical periods. This area has been declared a buffer zone for the World Heritage site.

The most critical value for preserving the cultural landscape is the view over the plains of the Jezreel Valley. This extensive landscape should be preserved as an open buffer between the existing and anticipated construction between Afula and Tel Megiddo. Preserving the area defined in the buffer zone ensures that it will be possible to stand atop the *tel* and continue to imagine the past wars and the future war, stated in the Book of Revelation, over the natural topography unmarred by construction.

Today the early Christian chapel is located within the compound of the Megiddo Prison. However, the chapel was incorporated into the cultural landscape polygon in the hopes that future changes in the prison structure could create a functional and scenic link between the chapel, the Kini stream, Tel Megiddo and the plains where the future battle of Armageddon is prophesied to take place.

Cultural Anchors

The summit of Tel Megiddo, the ancient prayer house.

Fig. 5.11 The Bahá'í Gardens—Al Bahaja

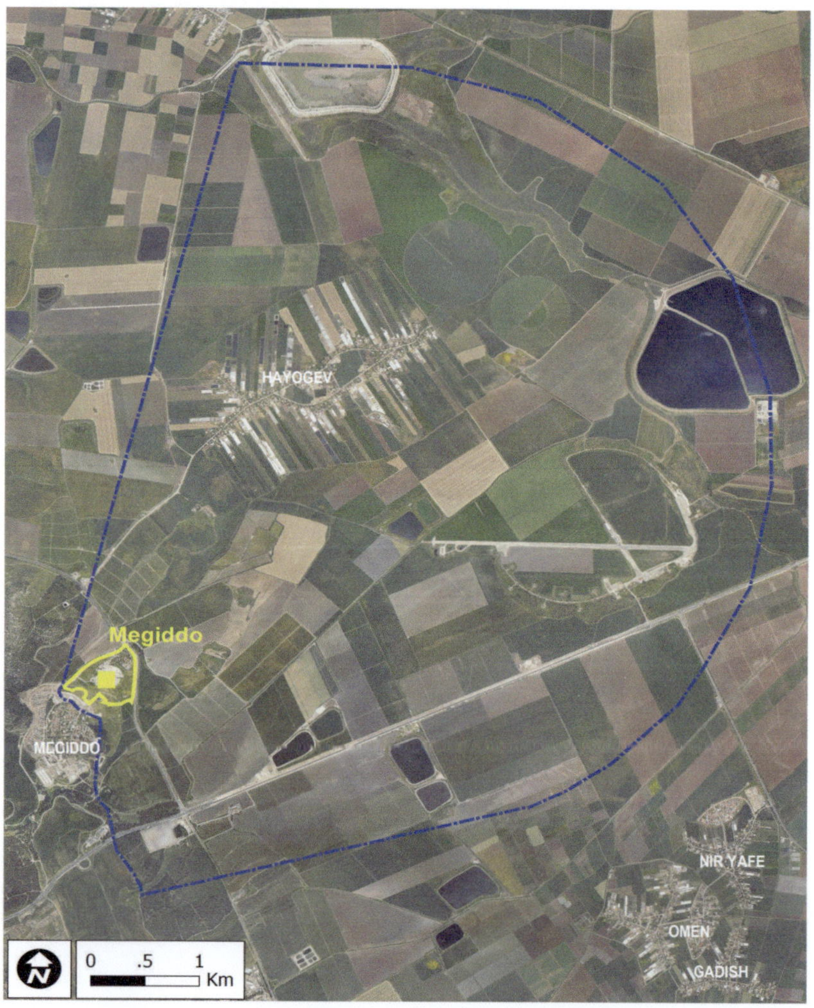

Fig. 5.12 Megiddo and Armageddon—polygon

Fig. 5.13 View from Tel Megiddo

5.1.2.3 "Land of the Caves": A Relict Organically Evolved Landscape as Part of the Caves of Maresha and Beit-Guvrin in the Judean Lowlands as a Microcosm of the Land of the Caves (Figs. 5.15, 5.16 and 5.17)

Narrative

The "Land of the Caves" in the Judean Lowlands was declared a World Heritage Site in 2014. While caves are present in other regions in Israel, including the Galilee and the Carmel, they are natural caves found in limestone. The "Land of the Caves," an area of refuge caves and hiding complexes, has many thousands of caves carved throughout the region by quarrying the soft chalk stone characteristic of the Judean Lowlands.

The plain maintained continuous and relatively dense habitation throughout different historical periods, primarily based on agriculture. These communities left their mark on the landscape with the remains of settlements. The most significant remains are the cities of Maresha and Beit Guvrin, along with the ruins of the villages of Madras, Burgin, Atari, Refa and others. Under the remains of these settlements, there is a relatively large concentration of underground complexes. Further away, the caves become increasingly scattered, and their number decreases. The caves inside these communities were connected to the structures above them. They were initially used as a quarry to produce building materials to construct the settlements. Later they were utilized for domestic and other agricultural purposes.

Three types of quarrying have been identified, associated with different periods:

Underground facilities—relatively small caves from the Hellenistic period, primarily between 300 and 200 BCE. This form of the cave quarries corresponded to the functions defined for them in advance, following the extraction of stone used for building materials. For example, the contours of a cistern were rounded; an olive press was square, and a columbarium was formed as a double-cross. These caves, concentrated primarily around Maresha, formed a "cellar" for the houses above them.

Hiding complexes—the hiding complexes from the Roman period constitute a special type of underground system. Studies have linked them to the Bar Kokhba revolt

Fig. 5.14 Tel Megiddo and Armageddon

(132–135 CE) and the period of preparations that preceded the uprising. Most of the systems were installed in existing underground systems linked by tunnels or narrow corridors. Only a few of these systems were created for hiding. There are dozens of such sites around the Judean Lowlands intended as hiding places for the population and perhaps also as a protected point for sorties attacking Roman soldiers.

The Roman historian Dio Cassius, a contemporary of that period, describes the phenomenon in his writings (*The History of Rome*, Volume 8, Book 69, paragraph 12):

> ...when Hadrian was away, they openly revolted. To be sure, they did not dare try conclusions with the Romans in the open field, but they occupied the advantageous positions in the country and strengthened them with mines and walls, in order that they might have places of refuge whenever they should be hard pressed and might meet together unobserved underground; and they pierced these subterranean passages from above at intervals to let in air and light.

Bell caves—these were large caves formed at a later stage. These caves are mainly found in Beit Guvrin, Lutzit and Amatzia. They were dug as pits, and the format included a small opening at the top in the calcrete rock, which widens into a bell shape, hence their name. These caves were used as quarries that supplied building materials to the coastal towns and Beit Guvrin, as described by Al-Muqaddasi, the famous Arab historian of the tenth century CE, "She (Beit Guvrin) is the land of wealth and abundance, containing many marble quarries…".

Quarrying bell caves peaked during the early Islamic period, between 700 and 1100 CE.

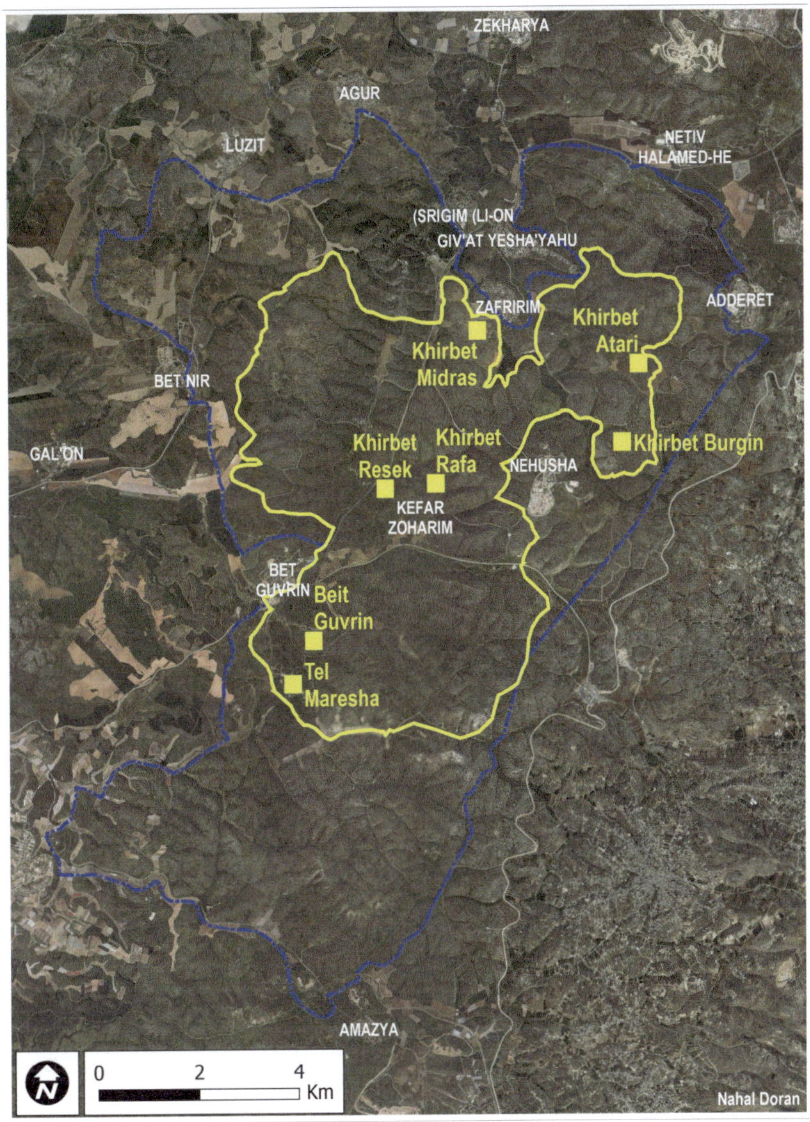

Fig. 5.15 Land of the Caves—polygon

The Landscape

The combination of the relief of the landscape, the soil and the climate make this area one of the most beautiful regions in Israel. The Judean Lowlands are characterized by rolling hills and relatively wide streams. Its morphology was formed by the erosion of the relatively soft underlying chalk and marlstone. Above the chalk layers is a layer of calcrete (locally called *nari*), a harder rock akin to limestone. The soil that develops on the *nari* and the chalk is rendzina, providing a dark contrast to the *nari* and the pale exposed chalk. At the base of the hills, the soil is dark brown and silty. There is very little tree growth on the slopes and valleys that lack the *nari* layer, and the diverse Mediterranean scrub that covers the hills occasionally alternates with green agricultural plots in the valleys. The Judean Lowlands lie on the transition zone between the Mediterranean and Irano-Turanean biomes and are located on the southernmost border of Mediterranean scrub. In the southern parts of the plain, there is less rainfall, and as a result, the flora is increasingly sparse as one travels south, transitioning from woodland to garrigue to dwarf-shrub steppe.

Fig. 5.16 Typical landscape in the region of the Land of the Caves

The natural landscape in the Land of the Caves has been preserved primarily due to the lack of modern settlement in the region. The cultural landscape of the Land of the Caves is hidden by vegetation. The entrances to the collapsed caves have a natural appearance because of eroded rock, patina and vegetation. These subterranean systems reflect the lives of the people who lived there and preserve the spirit of the area.

Defining Polygons

The section of the Judean Lowlands chosen to represent the phenomena of the caves is an area that is part of the Tzora geological formation in the Maresha area, whose chalk composition is particularly suited for quarrying. Caves and caverns are also found in Adulam, an area of over 51 square kilometers in which thousands of caves exist. The phenomenon of thousands of ancient caves carved into such a condensed geographic space is exceptional, both in quantity and in the diversity of cave types.

A buffer zone of 126 square kilometers surrounds this polygon which also contains quarried caves, although they are fewer in number and less diverse.

Cultural Anchors

Tel Maresha—an ancient city from the First Temple and the Hellenistic periods. The city contained approximately 160 cave systems for various purposes utilized by the inhabitants living in the houses above them. In a later period, hiding systems were hewn to connect the different spaces.

Beit Guvrin—an ancient town dating to the Roman and Byzantine periods containing dozens of caves, the most famous of which are the bell caves. These caves were used as stone quarries at the end of the Byzantine period and the beginning of the early Islamic period.

Khirbet Madras—a large village that was a Jewish settlement during the Second Temple and Roman periods, containing dozens of caves, including burial caves, columbaria and hiding complexes.

Fig. 5.17 A columbarium in the area of the Land of the Caves

Khirbet Atari—a small village associated with the Atari Village destroyed by the Romans during the first uprising. It was a Jewish settlement between the Hasmonean period and the days of Bar Kokhba. Dozens of caves were uncovered, containing cisterns, ritual baths, hiding systems and catacombs.

Khirbet Burgin—a large village with a Jewish population dating to the First Temple period. It was a possible site for Kfar Bish, a place mentioned during the first rebellion as a village that surrendered to the Romans. The site contains dozens of caves, many of which are burial caves, hiding systems and cisterns.

Khirbet Refa—a large Jewish village dating back to the Second Temple period destroyed at the end of the second century CE, in the Bar Kokhba Rebellion. All the cave typologies in the Judean Lowlands were found here, including cisterns, ritual baths, columbaria, workshops, stables and burial caves. At a later period, tunnels and connective spaces were added to create complex hiding systems.

Khirbet Resek—a hiding complex carved out of large bell-shaped pits at the foot of Tel Godoy, some of which were used as a columbarium.

5.1.3 Recorded on the World Heritage Tentative List

5.1.3.1 Christian Sites in the Galilee (Jesus and the Apostles): An Associative Landscape as Part of the Galilee Journeys of Jesus and the Apostles; and the Sea of Galilee and its Ancient Sites (Figs. 5.18, 5.19, 5.20 and 5.21)

Narrative

For the Christian pilgrim, the Lower Galilee region and the Sea of Galilee are the "Land of the Gospel." Nazareth is the site of the Annunciation. In the Galilee, Jesus delivered his sermons, wandered the paths and gathered his disciples, who spread the message of Christianity throughout the ancient world. The natural areas, including the olive groves, wheat fields, the sea and isolated hills, are the landscapes of the life of Jesus. Many Christians, scholars, clerics and pilgrims see this landscape as a "Fifth Gospel," in addition to the four New Testament gospels. The gospel personifies and shapes the events described in the texts; viewing the story through

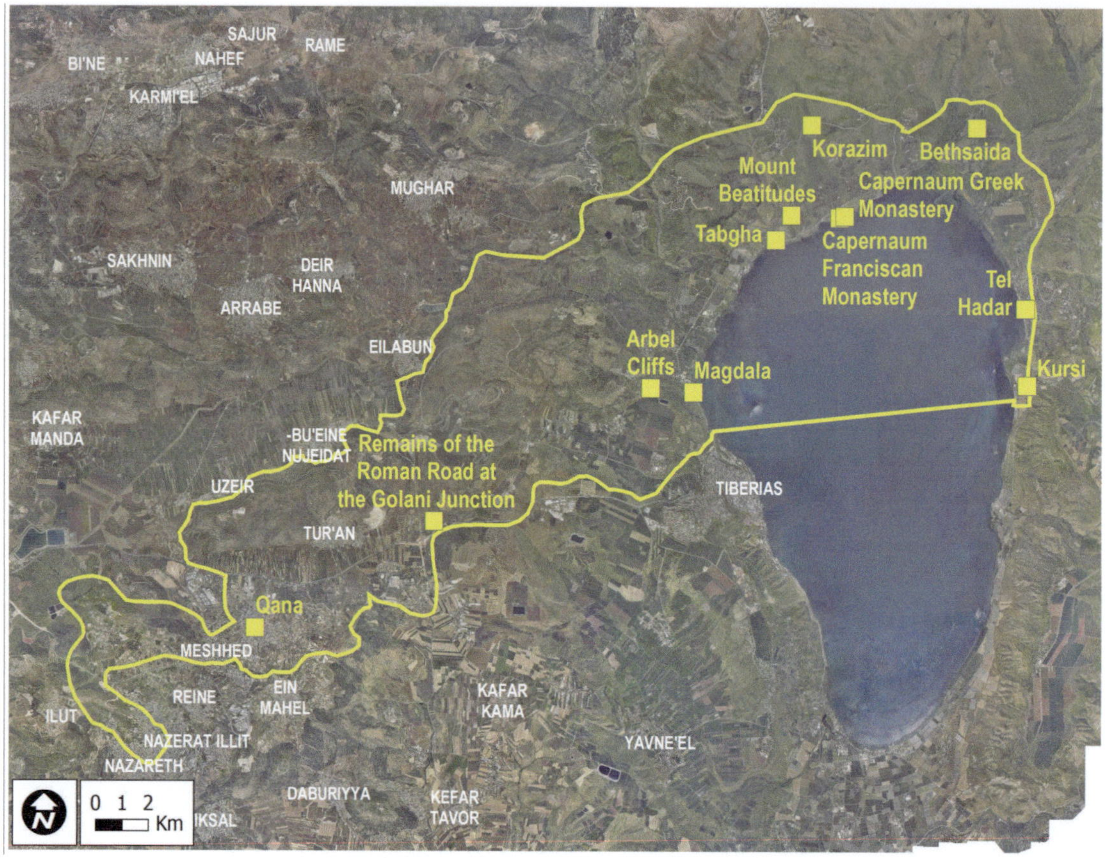

Fig. 5.18 Christian sites in the Galilee (Jesus and the Apostles)—polygon

Fig. 5.19 Church of the Beatitudes

Fig. 5.20 The eastern shore of the Sea of Galilee

Fig. 5.21 View of the Christian sites in the Galilee (Jesus and the Apostles)

the landscape provides a clearly defined backdrop to those gazing at the contours of the land. The Galilee landscapes are interwoven into the New Testament as places and sites where significant events occurred and in parables and sermons told by Jesus and the Apostles. Appreciating the landscape is vital to understanding the evangelists' words. It is difficult to picture the "Parable of the Sower"—perhaps one of the most important of Jesus's parables—without understanding the natural and agricultural landscapes of the Galilee. It is similarly difficult to envision the "Parable of the Blind" without seeing a spring in the Galilee in its natural setting. While most visitors to these sites are Christian pilgrims, the geographical landscape allows all visitors to imagine these events. The importance of the Fifth Gospel is not only religious; it also has immense cultural significance.

The Lower Galilee region and the shores of the Sea of Galilee still retain a composition of sites and landscapes that allows the viewer to imagine life during the time of Jesus. The olive groves and pomegranate orchards surrounding the Qana village today are similar to the agricultural environment where Jesus performed his first miracle. The Roman Road from Nazareth to the Galilee is not a holy place, but it allows the visitor to sense an important landscape of the past. On the coast of the Sea of Galilee, between Tabgha and Capernaum, outside the archaeological sites and the modern churches, is a small bay called "Bay of the Parables." This bay creates a natural theater and allows the visitor to imagine Jesus delivering a sermon, interspersed with parables, to a large crowd.

Defining Polygons

The polygons of the cultural landscape are defined to include the majority of the sites that have associative meaning from the perspective of the visitor observing the landscape. A substantial buffer was established that includes many of the sites mentioned in the Gospels describing the journeys of Jesus and the Apostles. The places that illustrate the stories are protected mainly as sacred sites and are run by one of the Christian denominations. These sites are either declared archaeological sites or recognized National Park sites. Many of these places are defined as anchors.

One can find grander, more beautiful churches throughout Israel and the world than those in the Lower Galilee and the Sea of Galilee. Indeed, Roman and Byzantine archaeological sites are more monumental and better preserved than the Galilean ones. Yet the value of this area is far greater than the sum of its parts. The purpose of defining polygons is to maintain its synergistic value. Otherwise, the hills of Nazareth risk being transformed into a large nondescript urban mass, and the shoreline of the Sea of Galilee could become merely a resort peddling beer and sunshine. The polygon does not include the city of Nazareth, as the city's current landscape has little connection to the image of the village where Jesus spent his childhood and youth. It does not include Mount Tabor, as the New Testament emphasizes the site's isolation and remoteness from the primary activities of Jesus and the Apostles. It does include the areas where Jesus spent most of his adult life since he left Nazareth until his final pilgrimage to Jerusalem, where he was executed by the Romans. The waters of the Sea of Galilee also have cultural heritage value, and the polygon includes a section of the sea. The sea is the medium that connects dozens of ancient anchor points and small harbors scattered along the shore. During the time of Jesus and the Apostles, the sea was the primary route connecting these points, and important miracles occurred on the water.

Cultural Anchors

Qana—since the Byzantine period, this village has been associated with the place where Jesus turned water into wine. Four churches were established during the nineteenth century to commemorate this miraculous event, the most prominent being the Franciscan Wedding Church and the Greek Orthodox Church.

Remains of the Roman Road at the Golani Junction—visible remains of a paved Roman road leading from Acre to Damascus (known as *Astrata*) are exposed north of the Golani Junction. This is the route that Jesus and his disciples likely took on their journey from Nazareth to the northern shore of the Sea of Galilee.

Arbel Cliffs—a vantage point in the Galilee region that provides spectacular views of all of the sites where Jesus was active around the Sea of Galilee and the Christian sites in the region.

Magdala—the central Jewish town along the Sea of Galilee during the first century. The town was known for its fishing and fish industry and was called *Magdala Nunaya*, the Tower of Fish. It is the hometown of Mary Magdalene who became one of Jesus's disciples. The few visible remains do not include any remains from the first century CE.

Tabgha—the Church of the Multiplication is a modern Catholic church built on the remains of a Byzantine church excavated in 1932. A beautiful mosaic floor from the fifth century CE was partially restored and rebuilt. Next to the stone below the altar, an ancient mosaic was discovered depicting a basket with bread and fish. According to Christian tradition, this is the site where the miracle of the Feeding of the Five Thousand took place.

Capernaum—this site played a significant role in Jesus's life. This is where he found the first Apostles, taught in the synagogue, astounded those who heard his sermons and performed many miracles. The town served as a base for journeys near and far; at their end, Jesus would return to Capernaum. The Catholic Church owns most of the land, including the ancient village. The remains include a synagogue and houses surrounding it. One of the houses was identified as Peter's house during the Byzantine period, and a large church was built on top of it. At the end of the twentieth century, a modern church that seemed to hover above the Byzantine church was constructed. To the east of the remains of Capernaum is the Church of the Apostles, a Greek Orthodox Church. Excavations have uncovered a Roman bathhouse from the second and third centuries CE, and below the bathhouse are the remains of an earlier structure from the time of Jesus.

Korazim—one of three cities where Jesus performed "great miracles." The remains of the town Chorazin from the Byzantine period include a fourth-century synagogue in which the "Cathedra Moses" seat was found, a term mentioned in the New Testament, a ritual bath, houses, industry, commerce and more.

Mount of Beatitudes—a mountain in northern Israel where Jesus is believed to have delivered the Sermon on the Mount. This site, next to Capernaum, has been commemorated since Byzantine times. A modern church was built on top of the mountain with magnificent views of the holy places attributed to Jesus and the Apostles.

Bethsaida—an archaeological *tel* in Jordan Park contains the remains of the town of Bethsaida, mentioned many times in the New Testament. The town Bethsaida, the biblical *Geshur*, is the birthplace of five (and some say six) of the twelve Apostles. The surrounding area is full of events recorded in the New Testament and includes places where important miracles occurred, including the miraculous healing of the blind. The excavations uncovered the first-century CE houses of a fisherman and a winegrower. In addition, notable remains from the tenth century BCE were discovered, including a magnificent palace, walls and an impressive gate.

Kursi—the remains of a town, port and monastery along the shores of the Sea of Galilee. According to the Christian tradition, the large Byzantine monastery was likely built where the Miracle of the Swine took place.

Tel Hadar—an archaeological *tel* on the shores of the Sea of Galilee between Bethsaida and Kursi. A relatively new tradition states that this is where the miracle of the Feeding of the Four Thousand took place, also known as the "miracle of feeding the nation."

5.1.3.2 Mount Tabor: An Associative Landscape as Part of the Galilee Journeys of Jesus and the Apostles (Figs. 5.22, 5.23 and 5.24)

Narrative

Residents living around Mount Tabor have always considered it to be holy. Some believe that even the Canaanite temples in Megiddo were built to face the distant mountain. During the biblical period, when the Israelite tribes settled the land, Mount Tabor marked the border between the tribes of Issachar and Naphtali. The foot of the mountain served as a gathering point for the armies of Israel in later periods (Judges 4: 6), and not only because of the mountain's strategic location. Barak, the Israelite military commander, led his army to confront the forces of Sisera, together with Deborah, the prophetess. Another battle took place between the Israelites and the Midianites (Judges 8: 18). Most Christian denominations believe that Mount Tabor was the site of the transfiguration when Jesus changed before the eyes of Peter, James and John: "His face shone like the sun and his clothes were white as light" (Matthew 17: 2). Pilgrims have flocked to the mountain since the fourth century and continue this tradition until the present.

The Landscape

Mount Tabor is visible over a great distance due to its isolation in the landscape, its signature domed shape and its height relative to its surroundings (588 meters above sea level, 450 meters above the Jezreel Valley). The mountain is an excellent example of a geological "horst," an area of raised block surrounded by subsided areas. Mount Tabor is situated between the rifts of the Nazareth Hills and the Tabor River and is covered with Mediterranean scrub, particularly at its peak and on its northern and western slopes.

A marked hiking trail encircles the top of the mountain, crossing a pine forest and Mediterranean woodland dominated by Palestine oaks (*Quercus calliprinos*) and the Mount Tabor oak (*Quercus ithaburensis*), offering spectacular scenic views of the Galilee. The Israel Trail also crosses the area, ascending to the top of the mountain and descending on the other side. A paved narrow winding road also reaches the summit. The modern road follows the ancient pilgrim route that includes more than 4300 carved steps.

A fortress from the Hellenistic period is located on the summit. In the first century CE, Josephus made improvements to the fortifications and created a large complex encompassing an area of approximately 40 ha. Tancred the Crusader, a Prince of Galilee, established a castle in the middle of the area fortified by Josephus. The

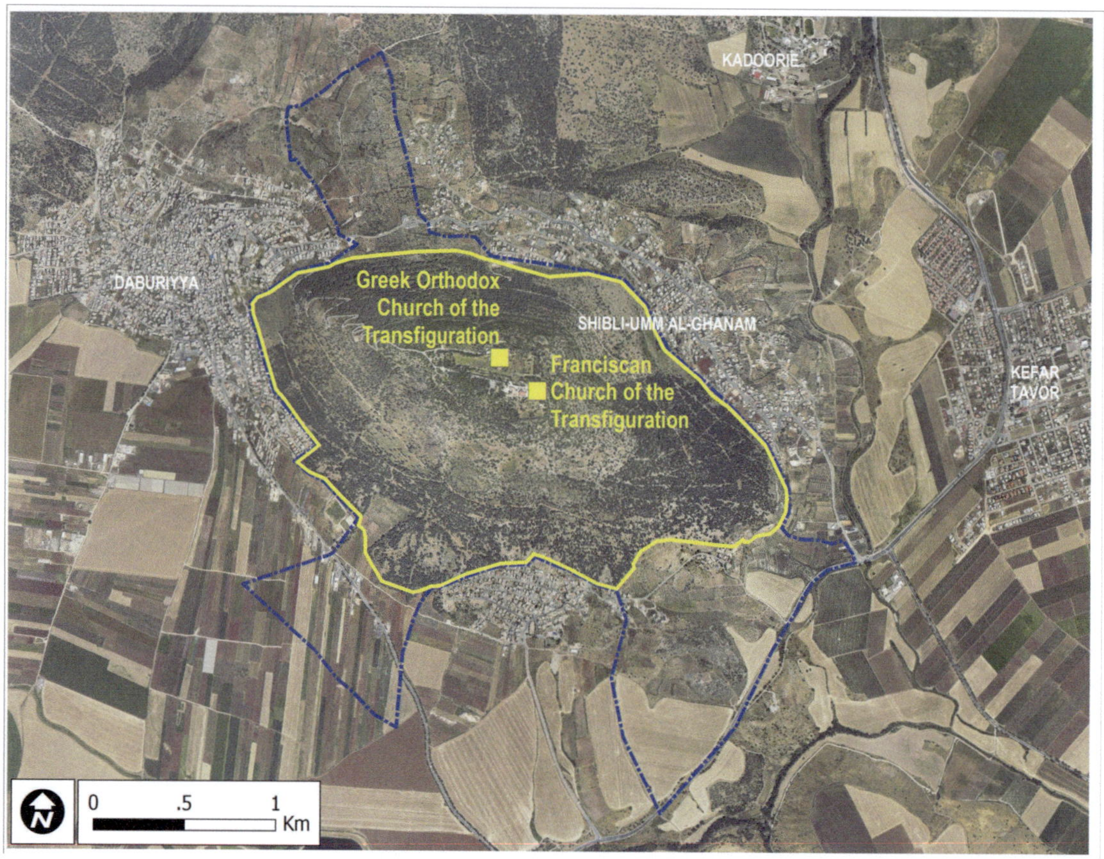

Fig. 5.22 Mount Tabor—polygon

different archaeological strata are evident to visitors. In 1873, the Franciscans built a small monastery at the top of the mountain that has significantly expanded over the years. A basilica completed in 1925 crowns the site. It was designed by the famous architects Antonio and Jules Barluzzi based on the plan of Mar Simian, a Byzantine church near Antioch. In the church crypt, it is possible to see the foundations of a former Byzantine church. The monastery, which is highly visible from afar, contains a guesthouse for pilgrims.

Near the Franciscan complex is a Greek church named for the prophet Elijah, which was built on the foundations of former Crusader and Byzantine churches that also bore Elijah's name. The Greek Church is next to the Cave of Melchizedek Chapel, which memorializes the tradition that Melchizedek, King of Salem, was a hermit who met Abraham on this site.

Defining Polygons

The cultural landscape polygon of Mount Tabor includes the entire mountain, except for the built-up area at the base of the village, whose construction has recently also expanded up the mountain. Mount Tabor should remain without any further construction on its slopes to ensure its continuation as an arena that visually contains the stories of Jesus of Nazareth and the stories of the fortifications during the Second Temple period, the Byzantine monks and the refortification during the Crusader period. The ascent to Mount Tabor should remain on the existing slow and narrow road so that future pilgrims may have a similar experience to the pilgrims of the past.

The villages built at the foot of Mount Tabor—Shibli, Daburiyya and Umm al A'nam—have expanded to form a continuous ring around the mountain. The buffer zone polygon of Mount Tabor defines corridors that connect the open area of Mount Tabor with the surrounding open space; a narrow space between Daburiyya and Shibli linking Tabor with the Beit Keshet Forest; a narrow strip that separates Daburiyya from Umm al A'nam and connects Tabor with the Jezreel Valley; and another narrow strip separating Shibli from Um al A'nam that connects the mountain and the stream. It is also recommended that all plans for construction or development in the buffer zone be reviewed within the context of maintaining the cultural landscape of Mount Tabor in relation to the surrounding Galilean landscapes.

Cultural Anchors

Franciscan Church of the Transfiguration, Greek Orthodox Church of the Transfiguration.

Fig. 5.23 Mount Tabor—Church of the Transfiguration

Fig. 5.24 Mount Tabor

5.1.3.3 The Horns of Hattin: An Associative Landscape as Part of Arbel (Arbel, Nabi Shu'ayb, Horns of Hattin)[1] (Figs. 5.25, 5.26, 5.27 and 5.28)

Narrative

On the July 4, 1187, the Frankish forces suffered a decisive defeat at the hands of the Muslims on the plain to the west of a volcanic hill known as the Horns of Hattin in the eastern Lower Galilee.

The Hattin campaign was part of the *jihad* war declared on the Latin Kingdom of Jerusalem some years before. The Muslims, led by Salah-ed-Din (hereafter Saladin), began gathering on the eastern side of the Jordan River in April 1187. On May 1, 1187, the Battle of Cresson took place after a large group of Muslim cavalries crossed into the Galilee for reasons that are unclear, and encountered a Frankish party led by the Templars and Hospitaller grand masters. The collision between the two forces by the springs of Cresson (Fontaine de Cresson/Croisson) was catastrophic for the Franks, with only the Templar grandmaster surviving the engagement a few of the parties' followers.

The Muslims reinvaded the Galilee in late June 1187. They crossed the Jordan River (June 27), captured the village of Kafr Sabt (June 30), and positioned themselves along other vantage points (such as Mount Tabor) in the eastern Lower Galilee and along the Safori-Tiberias Road. On the July 2, one of the Muslim forces conquered Tiberias and placed its castle under siege. Eventually, after a quarrel between the grandmaster of the Templar knights and Count Raymond of Tripoli, the Muslim provocation forced the Franks, led by King Guy of Lusignan, to evacuate their camps at the springs of Safori and to march on Tiberias. The Frankish forces started moving on July 3. When they reached the valley of Turan, the Muslims began attacking from their mounted archers. The march was slow and carried out at the pace of infantry, which likely encircled the three main bodies of mounted knights in rings of defense. According to the letter written by Saladin, the Franks managed to take over one of the water sources on their way (probably the spring of Turan) but later abandoned it. When the Franks reached Maskana, known for its large seasonal pool to the south, they were attacked, and the rear guard led by Balowin de Iblin retreated.

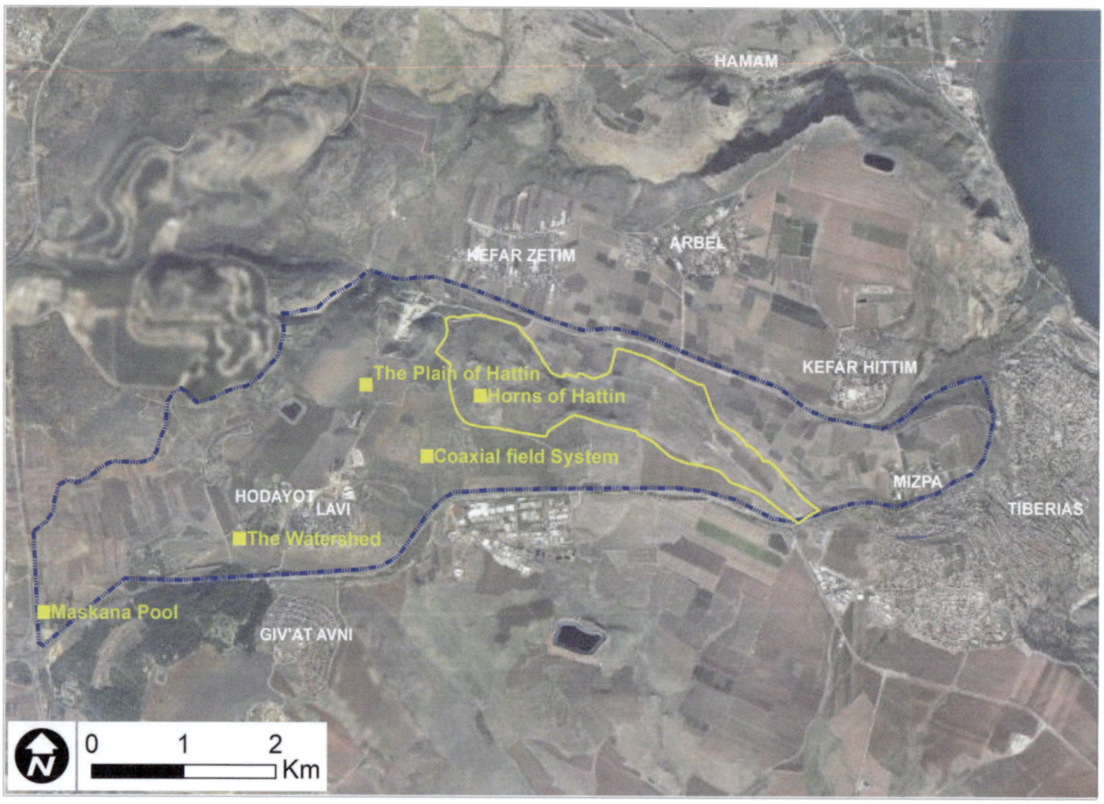

Fig. 5.25 The Horns of Hattin—polygon

[1] This section was written by Rafael Y. Lewis.

Fig. 5.26 Looking east from the Horns of Hattin

At sunset on July 3, 1187, the fighting had not stopped. The Franks camped around Maskana, while the Muslims brought convoys of provisions to the front line and at the same time, continued to attack the Frankish camp. At sunrise, the Franks were on the move again, the men and their horses exhausted. Some scholars have argued that Count Raymond of Tripoli took a northern route leading towards the Springs of Hattin in order to quench their thirst. Other researchers say that on July 4, the Franks were still heading toward Tiberias as planned. The exact order of events is unclear based on the different textual sources, but we know that the Muslims burned the fields surrounding the Franks at a certain point. At another point, the vanguard led by Raymond of Tripoli received the command (and honor) to charge against the Muslim troops blocking their way. But instead of the two forces colliding, the Muslims opened their ranks, and Raymond's troops went through them, never returning to the battlefield. It is also unclear when the infantry fled to the Horns of Hattin, leaving the main body unprotected, and at which point cavalry charges were made under the direction of Saladin. The Muslims' victory was marked by the capture of the relict of the "True Cross," the fall of the Frankish king's red tent, his imprisonment, the massacre of the Templar and Hospital knights, and the beheading of Count Reynald de Chatillion. The defeat of the Franks led to the fall of the holy city and the Latin Kingdom of Jerusalem.

The detailed landscape archaeology study done during the Battle of Hattin Archaeological Project, which involved the reconstruction of the Hattin landscape, allowed us, for the first time, to date the events outlined above chronologically and to locate them in the landscape. The 1187 campaign was not the only time when the landscape of Hattin became an arena of human conflict. The landscape of Hattin

Fig. 5.27 The Horns of Hattin in summer

played a role in the two Jewish revolts against Rome (66–70 CE and 132–136 CE), two of the battles of Napoleon's campaign in the Galilee (April 1799), a battle between Bedouin and Ottoman cavalry (March 1857), and the Israeli War of Independence War (July 1948). All of these conflicts left a material imprint on the Hattin landscape.

The Landscape

The Battle of Hattin is named after the volcano of Hattin, a dominant feature in the landscape. The decisive stage of the battle took place at its foot. Nevertheless, we should point out that the mountain itself did not play any significant role in the major event described. It was one of the thousands of natural and artificial features in the Hattin landscape at the time of the event. These features, which include water sources, roads, paths and field systems, stone boundary walls and rural terraces, may be considered less impressive than the volcano, though together, they played a significant role in the historical event of July 1187. The Hattin story is only understandable when these landscape features are considered.

The Battle of Hattin was an open field battle: the Muslim and Frankish forces fought against each other immediately after the Franks left the springs of Safori and began marching towards Tiberias. This means that, at least theoretically, the immediate area of engagement extended over 30 square kilometers. This area has seen extensive modern development; it has been cultivated intensively, modern roads have been paved, and new villages have been founded.

Though some would suggest that the decision to leave Safori was the actual cause of the Frankish defeat, the landscape, which played a decisive role in the battle, is much smaller than previously presented. The "story" of Hattin can be described as follows: the landscape extends west to Maskana (Golani Junction); east to the eastern slopes of the Horns of Hattin; south to the area between the ruins of the abandoned village of Lubia, Golani-Tiberias Road (No. 77) and the abandoned village of Nimrin; and to the northern slopes of the Hattin plain to the north. The area demarcated was the arena of the final and decisive stages of the battle, and its landscape reflects the events between the night of July 3 to midday of July 4, 1187.

Fig. 5.28 The Horns of Hattin

The reasons for choosing this area for the study were entirely pragmatic. First, the fact that the area was still relatively undeveloped made it possible to demonstrate to the modern visitor a landscape very similar to what the Muslims and Franks would have experienced at the time of the actual battle. Second, the landscape of Hattin is a good case study for studying a landscape of conflict since its bloody meadows have witnessed many other conflicts and violent clashes during different periods in time. Despite the widely different technological warfare capabilities developed over time, these events undoubtedly shared some common characteristics.

Over the years, the Hattin landscape has been the focus of a lot of human activity, though not necessarily of a violent character. The Hattin landscape has abundant archaeological sites, notably urban settlements (fortified and unfortified) dating to the Late Bronze age. Some Hattin urban sites have been continuously occupied for over 2000 years. An extensive road network was preserved in this small area, including two international highways, a public road, and abundant regional and local roads connecting various settlements along with their agricultural and industrial landscapes. A basalt quarry dating from the Mamluk period was found near one of those roads.

There are three kinds of relict fields in this landscape. First is the coaxial field system, a dominant feature in this landscape, dating to the Severan Emperors of the late second until the mid-third centuries CE. This system is very important for understanding the development of the Hattin landscape and its human influence. It undoubtedly influenced the mobility of armed troops and inevitably channeled the movement of troops onto the road systems. This had fatal implications on the historical events that took place in the area from the second century CE and had a particular impact on the famous battle of 1187.

The second relict field comprises the industrial and agricultural installations of the Hattin landscape, dating to different periods, including water sources composed of four seasonal pools and over 300 water cisterns, mainly concentrated within the settlements. Three inns (*khans*) from the Byzantine and Mamluk periods were positioned at the main road junctions. Burial sites such as rock-cut tombs and trench burials were used to bury the members of the local population and the casualties of the area's many conflicts. Stone piles were also a common phenomenon in the Hattin landscape. Some rock piles date to the Middle Paleolithic period, while others represent burials or agricultural stone clearances. There are two army encampments, one dating to the Battle of Hattin and the other to Napoleon's campaign in the Galilee in April 1799. Another interesting phenomenon in the Hattin landscape is the hideout complexes, though their dating is disputed. Some of these complexes were discovered outside the boundaries of settlements, at locations likely used as vantage points and strategic positions.

The third type is the religious landscape, with sites concentrated on the eastern side of the survey area, along the cliff overlooking the Arbel Valley. These include sites such as the Horns of Hattin (identified as the Christian Mount of Beatitudes); Nabi Shu'ayb (the Druze religious shrine and traditional burial place of biblical Jethro); Khirbet Kankuza (an Iron Age IIa-b sanctuary with a standing *massebah* stone); and the area known as the "Stones of the Christians" (one of the traditional sites for the "Feeding of the Multitudes"). These features and many others give the Hattin landscape its distinct character, composed of numerous and diverse layers of natural and cultural characteristics.

The Hattin landscape also contains memorials. The remains of the dome of victory (*qubbat al-nasr*) built by Saladin on the southern peak of Mount Hattin after his triumph, can still be seen today. In other cases, fields were named after key figures and events. For instance, one field marked on a mandatory map from 1938 is called *Ard Maqtal en Nasara* (i.e., "the land of the massacre of the Christians").

Defining Polygons

The cultural landscape polygon extends from the Golani Junction in the west to the eastern slopes of the Horns of Hattin in the east and includes the Arbel Valley. It lies between the ruins of the village of Lubia, the Golani-Tiberias Road (No. 77) to the south, and the village of Nimrin and the northern slopes of the Hattin plain to the north, where the final and decisive stages of the Battle of Hattin took place. The main arena of the battle has been reduced to an area of 6.5 square kilometers from the larger area of 30 square kilometers. Still, recognizing the features that relate to significant parts of the Hattin story and protecting as much of this landscape as possible from modern development are crucial.

In recent decades, the Hattin landscape, like many other sites, has seen substantial modern development. However, contemporary visitors can still experience the landscape as it looked during the period when the Franks and Muslims clashed in the Battle of Hattin.

Cultural Anchors

The Maskana pool—a small Byzantine and Medieval settlement located to the north of the Golani junction, known for its small seasonal pool (46.5 meters in diameter) located to its south. The Franks camped in Maskana during the night of July 3, 1187.

The watershed—also known as the Hills of Lavi, a land feature that is part of the central watershed of Israel. The hills of Lavi run from the village of Lubia in the south, through Kibbutz Lavi, Hodayot and the village of Nimrin to the north. The watershed was the main land obstacle on the route the Franks took to Tiberias. This cluster of natural and artificial features influenced the outcome of the Battle of Hattin.

The Plain of Hattin—the fields extending between the Hills of Lavi and the Horns of Hattin where the final stages of the battle took place.

The oaxial field system—also known as El-Kasair ("the fragmented"). An area rich with relict fields, ancient roads, and agricultural and industrial installations. El-Kasair played a significant role in the Hattin story. This area is under serious threat due to current development plans.

The Horns of Hattin—a volcanic hill overlooking the Hattin plain and the Arbel Valley. This hill, with its two volcanic craters, was also a Late Bronze and Iron 2a-b settlement. The hill is the dominant land feature in the area. The final stages of the Battle of Hattin took place on the slopes and fields at its base.

5.1.3.4 Mount Karkom (Har Karkom): A Relict Organically Evolved Landscape (Figs. 5.29, 5.30 and 5.31)

Narrative

More than a hundred sites from the Stone Age have been discovered on Mount Karkom (Har Karkom), mostly from the Middle Stone Age (120,000 to 45,000 years ago). Archaeological surveys on the mountain have revealed numerous workshops for flint tools and indications that residents were living in *sukkot* (temporary outdoor shelters mentioned in the Bible and early texts). Settlement on the mountain also flourished at the end of the Chalcolithic period (ca. 3300 BCE), in the Early Bronze Age and at the beginning of the Late Bronze Age (ca. 2000 BCE). More than 200 sites relating to these periods were discovered on the mountain, including stone building remains, burial tumuli and stone monuments. One reason for these findings is the abundance of flint rocks on the slopes that remain in excellent condition. The mountain has gained international fame due to thousands of etched rocks in different styles at

Fig. 5.29 Mount Karkom — polygon

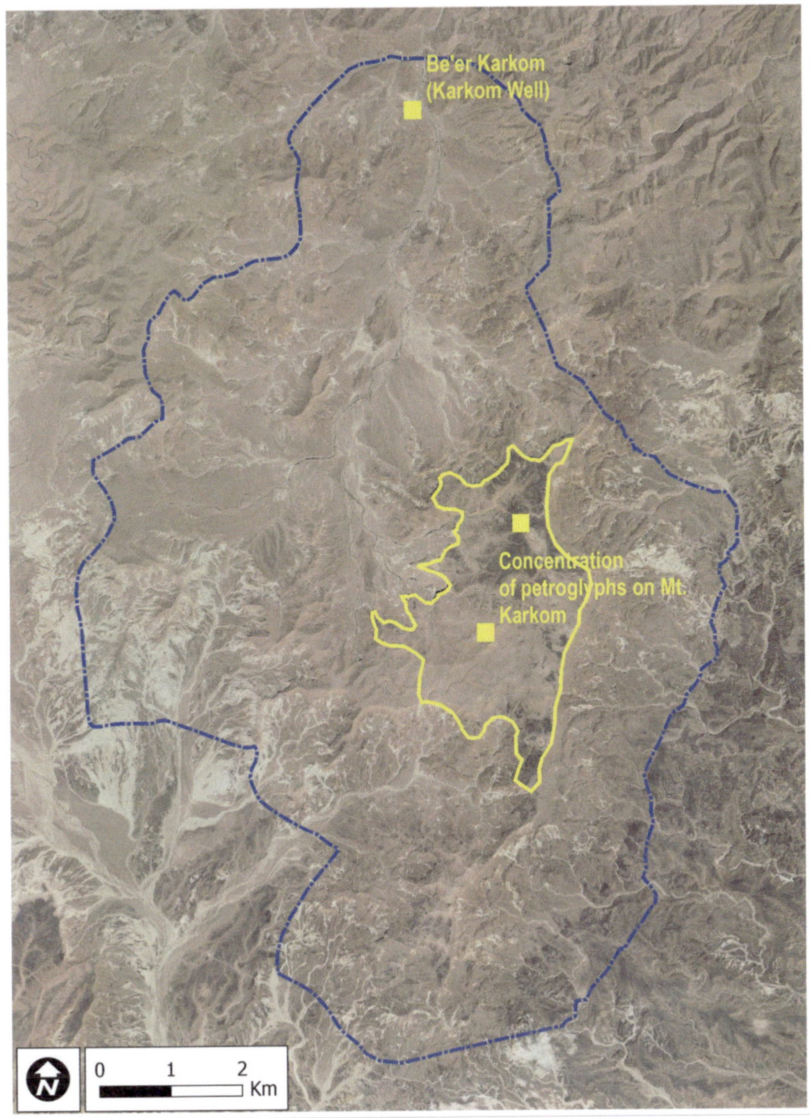

the site. The paintings on the rocks depict, among other things, animals and hunting scenes. Archaeologists believe that people who belonged to nomadic societies from these and later periods made these rock carvings. Some of the paintings from the Early Bronze Age indicate the domestication of animals. Prof. Emmanuel Anati, the excavating archaeologist, attributed ritual significance to the paintings, and in his opinion, Mount Karkom is associated with the biblical Mount Sinai. However, most scholars reject this view.

The Landscape

Mount Karkom is a large mountain in the Paran River Basin and is categorized as a tabletop mountain. The mountain peak is broad, flat, and composed of limestone, extending over seven square kilometers. In the past, the mountain ridge was part of a level plateau in the central Negev Mountains. Most of the plateau has been eroded by river floods. Several peaks remained, including Mount Karkom. The plateau weathered easily due to its soft composition of chalk and clay from the Maastricht and Paleocene periods, with only an Eocene upper shell made of hard limestone.

Fig. 5.30 Rock paintings on Mount Karkom

Fig. 5.31 The landscape of Mount Karkom

Trails lead up to the top of the mountain, and the flat top has an elaborate system of paths guiding hikers to interesting concentrations of rock paintings. Vistas atop the mountain offer spectacular views over the eastern mountains of the Sinai, the Arava, the mountains of Edom, the Paran River Basin and the northern Negev. The mountain is currently located within a military firing zone and as a result, it is only open to the public at specified times.

Defining Polygons

The cultural landscape polygon of Mount Karkom was defined as the top of "Table Mountain," an area where most of the remains of ancient human activity are concentrated. The buffer zone was defined as a polygon that connects the highest points surrounding Mount Karkom, creating a tableau with the surrounding mountains. The site is included on the Israel World Heritage Tentative List and could be extended to include the remains in the High Mountain in Jordan and the Sinai Peninsula in Egypt.

The polygon and the enveloping buffer zone are all located on the border of the Negev Highlands Nature Reserve, ensuring the preservation of the cultural landscape.

Cultural Anchors

Concentration of petroglyphs on Mount Karkom, Karkom Well (Be'er Karkom).

5.1.3.5 Timna: A Relict Organically Evolved Landscape (Figs. 5.32, 5.33, 5.34 and 5.35)

Narrative

Between layers of white sandstone, concentrations of copper were discovered in the Timna Valley in antiquity. Copper

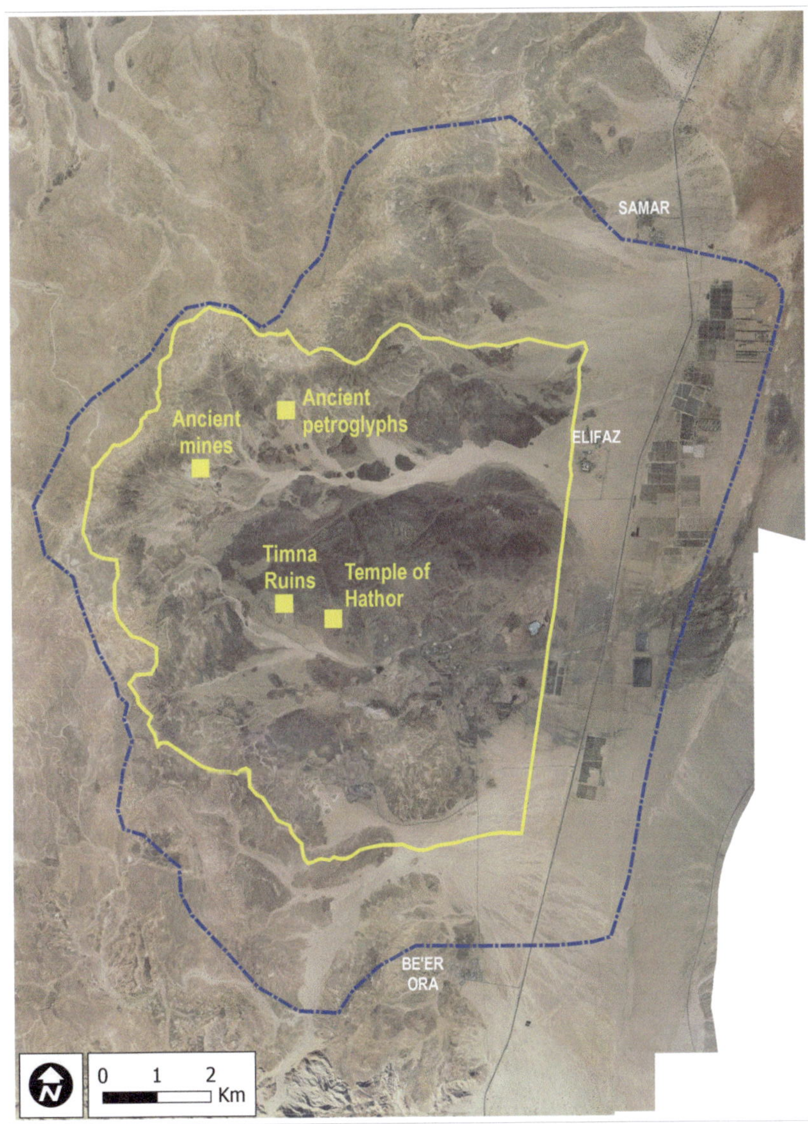

Fig. 5.32 Timna—polygon

Fig. 5.33 The remains of the copper mines at Timna

Fig. 5.34 An opening to the mine shaft at Timna

Fig. 5.35 Solomon's Pillars, Timna

extraction in the valley began in the late Stone Age (6th millennium BCE). During this period, copper was not smelted and used as a metal, however, the deposits were used to create jewelry. During the Chalcolithic period (4th millennium BCE), copper was mined in shallow mining tunnels. Mining peaked during the new Egyptian kingdom, 1400–1200 BCE. The Egyptians dug deep shafts to reach copper, the deepest of which was 37 meters. Copper metals were smelted and separated from the slag in the mining camp established in the middle of the valley, and the scattered remains are visible on the ground today.

The Landscape

The Timna Valley has several sandstone layers: red sandstone from the Paleozoic era at the base (the Solomon's Pillars formation), overlaid by a variety of sandstones, Middle white sandstone (Amir and Avrona formations) and a mixed sandstone layer from the early Cretaceous period. Following the stratification of the sandstone, a block of igneous rocks of granite and syenite rose up and created a fold structure. In the later stages of the crystallization of the magmatic rocks, veins formed containing rich copper compounds. Copper reached the white sandstone after the copper compounds dissolved in the source rocks, which were transported in solution and redeposited into the sandstone.

Mining tunnels, the remains of mining and smelting camps left by copper production operations, as well as other cultural remains of miners from different periods are evident to this day in the valley sandstone. These include a temple to the Egyptian goddess Hathor and rock paintings, such as Ramses III giving an offering and other rock paintings from that period, primarily depicting hunting scenes. It is possible to learn about the fauna that characterized the region for thousands of years, to identify tools and hunting tools and to know more about the lifestyle of the people who worked in the mines and traveled the local roads.

The Timna Valley is surrounded by cliffs that tower 300 meters above the valley floor, from the top of the Mil han Ascent to the north through Mount Berekh, Mount Gadna and Mount Hakhlil in the south. The valley opens towards the Arava on the eastern side and is reminiscent of the Crater in the Negev, although its geological formation was different.

Defining Polygons

Human activity related to the long history of copper mining is concentrated at the sandy base of the valley. The cultural landscape polygon boundaries were defined to reflect this story. Most of the cultural landscape area is currently a landscaped park with over twenty hiking trails and an artificial lake at its center.

The valley is surrounded by cliffs that clearly define the landscape, creating a sense of disconnection in time and space that contributes dramatically to the experience of the valley's cultural values. The arc of cliffs on the ridgeline was defined as the boundary of the buffer between the cultural landscape of the Timna Valley and its surroundings. The construction of buildings should be prevented in the natural buffer area of the Timna Valley, particularly structures such as utility poles and antennas that are visible from afar. South of the historic cultural landscape is a vast, disturbed landscape that contains the abandoned remains of a modern mine, with piles of earth and gaping pits. This area was not included in the Timna Valley cultural landscape, as it is not visible from the valley, yet it would be appropriate to restore the surrounding environment. The site is included on the Israel World Heritage Tentative List and could be extended as part of a serial nomination with the extensive mines at Feinan in Jordan, Timna's reflection in the Great Rift Valley.

Cultural Anchors

The Temple of Hathor, Timna ruins, ancient mines, ancient petroglyphs.

5.1.3.6 The Hula: A Continuing Organically Evolved Landscape as Part of the Great Rift Valley—Migratory Routes—The Hula (Figs. 5.36, 5.37 and 5.38)

Narrative

> *We will go up to the Hula and drain its swamps*
> *We will go up to the Galilee and we will rise and inherit the Hula.*

Teenagers enthusiastically sang this song after Yehoshua Hankin acquired a permit to drain the Hula Valley in 1934 from the British mandatory authorities. At the time, the valley contained approximately 3100 ha of marsh and a lake with an area of 1400 ha. Following the establishment of the State, the Jewish National Fund was tasked with draining the swamp. The drainage program had several objectives, including preventing malaria (although the disease had already been eradicated): land reclamation for agricultural use, water conservation (saving water that evaporated from the lake), and using peat for agricultural purposes.

The drainage project alarmed the local community of nature conservationists, as the swamp and lake contained abundant natural habitats. The area served as a stopping point for hundreds of thousands of migratory birds and was a haven for many species of birds that nest in Israel. Scientists and conservationists suggested establishing a "nature reserve" (though that term had not yet come into existence). The idea of preserving a swamp seemed unusual at the time. Eventually, the idea was accepted, and the first nature reserve in Israel was established in the valley in an area of approximately 320 ha.

There were high expectations for the peat at the bottom of the lake, yet they soon turned to disappointment. The peat had settled more than three meters below its original level. The only soil suitable for agriculture was at the bottom of the lake.

The heavy rains of the winter of 1991–1992 caused a large lake to form. The Jewish National Fund and the local kibbutzim who farmed the land decided to establish an ecological tourism project. They flooded 100 ha of the peaty area and called it "Hula Lake Park." The original Jordan River channel was deepened and water flowed into the lake before reaching the Sea of Galilee. The lake provides shelter for many species of birds and aquatic plants. In winter, more than 10,000 cranes stop there every night. Birdwatching observation points were set up around the lake, along with hiking trails and a state-of-the-art visitors' center, which developed guiding methods with minimal environmental impact. The Hula Nature Reserve is a fascinating place to observe birds. It is a declared wetland Ramsar site, and it is included on the Israel Tentative List for World Heritage.

Defining Polygons

The Hula Valley cultural landscape polygon includes the nature reserve and the Einan stream that feeds it. The polygon also includes the lake and the agricultural land between the drainage channels of the Hula Valley, called "the East and West Jordan Canals." This is the lowest area in the valley and the site of the original lake. The cultural landscape story focuses on the natural history of the valley, its drainage and reflooding. Around the polygon, an agricultural buffer was defined including the communities of Kibbutz Naot

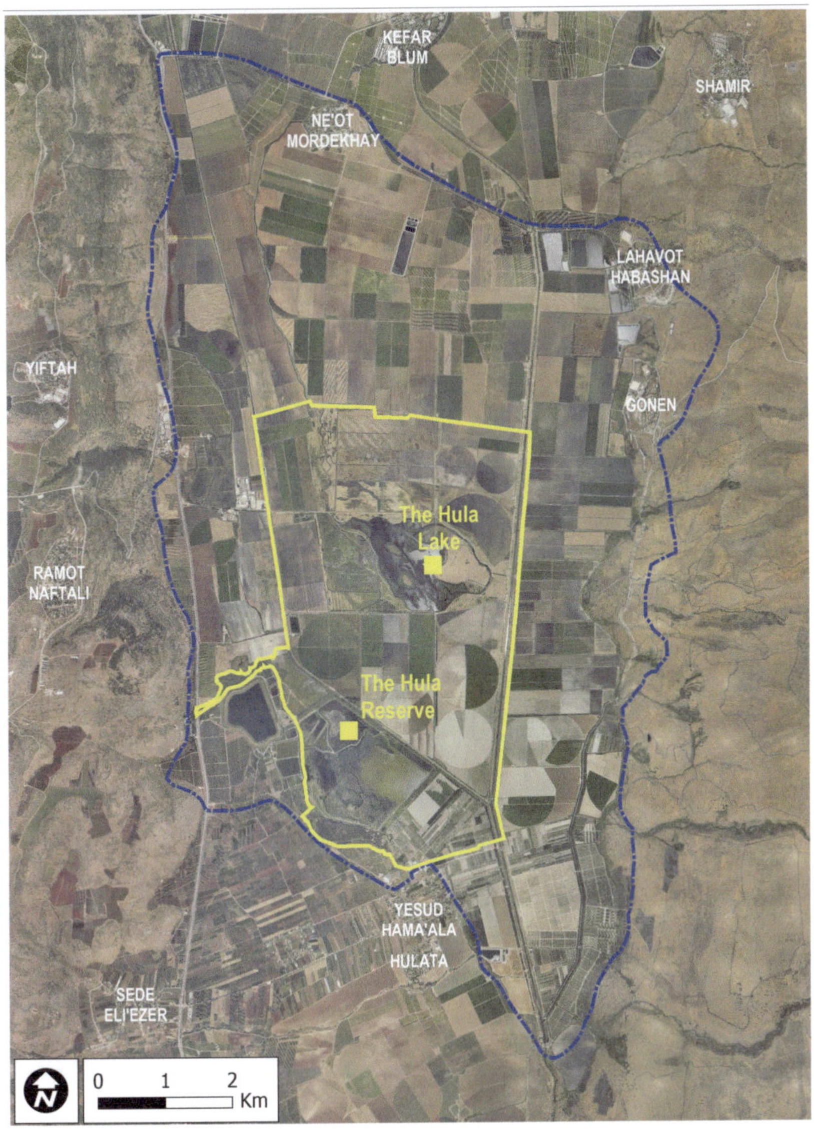

Fig. 5.36 The Hula—polygon

Mordechai, Lehavot HaBashan and Gonen, located at the edge of the area that had been flooded in the past.

The Landscape

The Hula Valley includes the agricultural landscape on land reclaimed from the drained marsh, the remains of the natural marsh, and the new lake, which attempts to recreate the original marsh landscape for visitors. Due to soil stability and drainage issues, large settlements and prominent buildings were not built in the center of the valley. The valley's landscapes, between the Golan Heights to the east and the Naftali Mountains to the west, have become a tourist center, attracting thousands of visitors each year who come to bird watch and enjoy the blend of agricultural fields, flowing water, and wide-open spaces. The sweeping landscape clearly "tells" a story of the interaction between man and nature in modern times.

Cultural Anchors

Hula Nature Reserve, Hula Lake.

Fig. 5.37 A view of the Hula Valley and the new Hula Lake

Fig. 5.38 The Hula Valley with a view of Mount Hermon in the distance

5.1.3.7 Nahalal: A Designed Landscape (Figs. 5.39, 5.40 and 5.41)

Narrative

In 1919, Eliezer Yaffe published a proposal to establish a new form of settlement: a hybrid between the private settlement of the moshava and the collective settlement of the kibbutz. The proposed settlement, later known as the moshav, merged the principles of social justice with the independence of the family unit and free enterprise. The moshav was an agricultural settlement based on mutual assistance and the cooperative marketing of agricultural produce.

At a Young Workers (HaPoel HaTzair) summit held at Kinneret in 1919, seventy senior workers adopted the idea. They established the first moshav in September 1921. They called the settlement Nahalal after the biblical city of Nahalal which was located on the border of the Tribe of Zebulun (Joshua 19: 15). The moshav was very successful and led to the establishment of many more such communities.

The Landscape

Architect Richard Kauffmann designed Nahalal in an iconic, circular form that reflected the community' unique ideology. Its public buildings were constructed on a hilltop that rose up in the center of the circle. These included a community center, school, a general store and a health clinic. The homes of the moshav members lined the road that encircled the central hill, and each house lot included a long field behind the house, forming a radial pattern. The disadvantage of the circular design was its closed form, making expansion impossible. As a result, the circular moshav became a one-time creation in the history of Zionist settlement. In 1922, Kauffmann designed the layout of Kfar Yehezkel with a design similar to that of Nahalal. However, he added a single straight road to the circular design to allow for future expansions.

Defining Polygons

The polygon for Nahalal follows the circular built area and the adjoining fields. Its original design remains to this day.

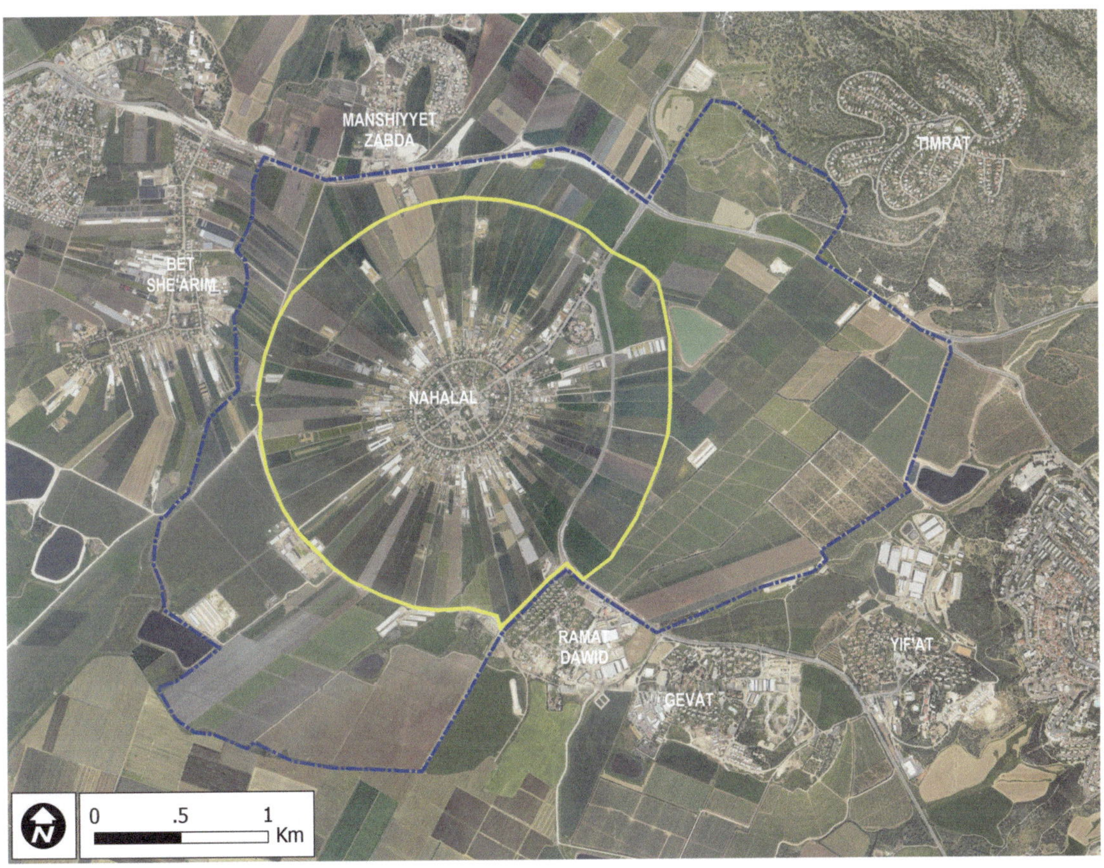

Fig. 5.39 Nahalal—polygon

Fig. 5.40 Nahalal

Fig. 5.41 The Nahalal commemorative stamp

All new construction in the moshav must maintain the unique form of the settlement. However, the integrity of the perfect circle has been harmed by the construction of Highway 73, which has cut into some of the plots. Additionally, building too close to the outer ring will likely damage the unique presence of the fields in the Jezreel Valley. A buffer zone was defined between the perimeter of the circle and the agricultural fields of the valley landscape outside of Nahalal to prevent massive construction adjacent to the designed landscape. The buffer area also includes the Nahalal cemetery in Tel Shimron, where many moshav members are buried.

5.1.3.8 Crusader Structures in the Western Galilee: A Relict Organically Evolved Landscape as Part of the Crusader Fortresses (Figs. 5.42, 5.43 and 5.44)

Narrative

The Crusader armies crossed the Sulam Ridge on May 24, 1099, and conquered Jerusalem in July of that year, thus consolidating their rule over most of the western Holy Land. The Principality of Galilee was one of the four major seigneuries of the Crusader Kingdom of Jerusalem. The Galilee itself was divided between the Principality of the Galilee, whose capital was Tiberias, and the Royal Estate of Acre. At this time, Joscelin de Courtenay III, Prince of the

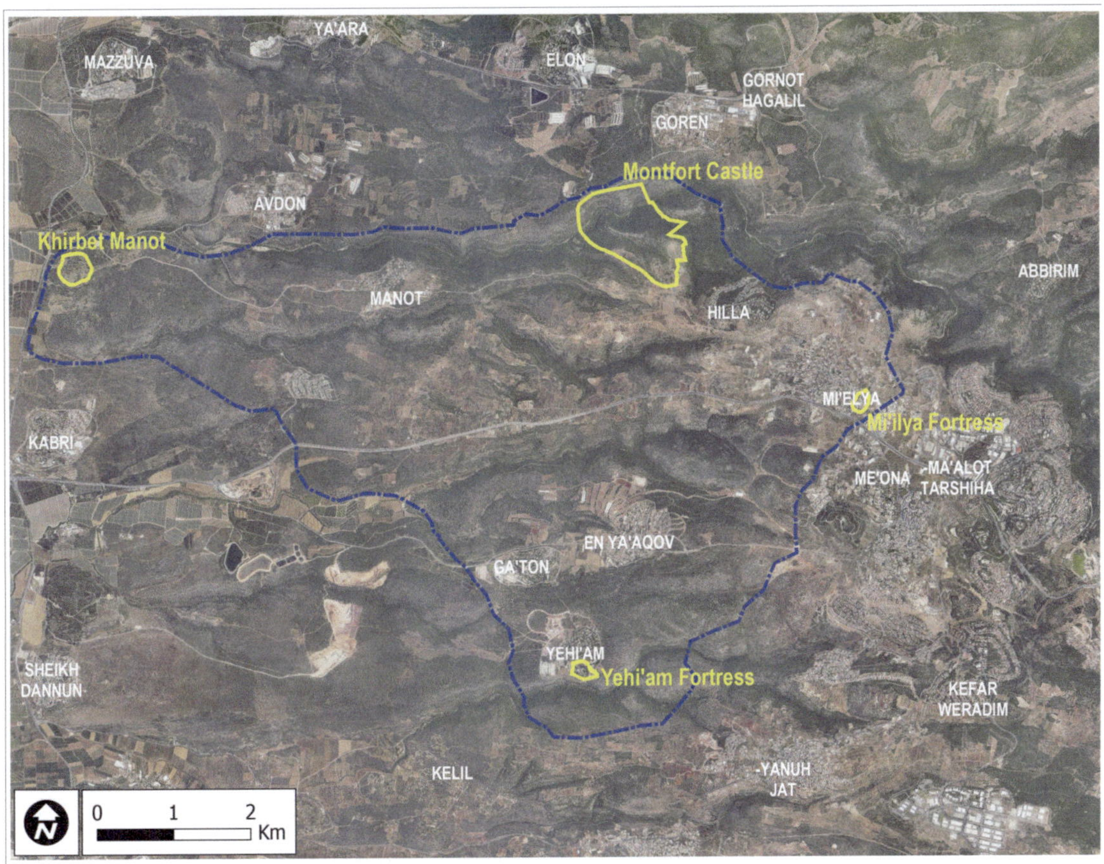

Fig. 5.42 Crusader structures in the Western Galilee—polygon

Galilee, established "the King's Castle" in Mi'ilya, and controlled most of the western Galilee. This period also saw the establishment of small feudal estates, such as Casale Umberti (Achziv) and the monastery named after St. George (Deir al-Assad).

After the Crusader defeat in the Battle of Hattin by Saladin (1187), the entire kingdom fell except Sidon (see Sect. 5.1.3.3). In 1191, Richard the Lionheart led the Third Crusade to regain Acre. After the conquest, Acre was declared the capital of the Crusader Kingdom, slowly expanding its hold over the western Galilee. In 1192, the border passed near Montfort. In 1229, the Crusaders expanded their reach to include Meron, and by 1240, their rule extended to the Jordan River. During this period, the military orders were the landowners who controlled the fortresses instead of the feudal knights. The Teutonic Order held the Montfort fortress, the King's Castle in Mi'ilya and the G'din Fortress (Yehiam). In the second half of the thirteenth century, the Muslim armies, under the command of Baybars, took control of the Crusader Kingdom and destroyed it. Over time, the farms and forts erected by the Crusaders became the nucleus of the villages established by the Muslims.

The Landscape

The Crusader period left an indelible mark on the western Galilee with its beautiful and impressive structures. The area with the highest concentration of Crusader buildings is now largely covered by dense Mediterranean scrub, where Arab villages were built. Some of these formed around the nucleus of Crusader farms or buildings.

The slopes of the western Galilee are etched with relatively steep streams. The remains of the Crusader fortifications can be recognized from a distance between these streams. The formerly agricultural areas between the

Fig. 5.43 The Yehiam Fortress

Crusader sites have largely reverted to woodland. Only a few small plots remain in cultivation.

Defining Polygons

The fortresses and other elevated vantage points, such as Mitzpe Goren, provide prospects of the landscape that allow visitors to view the past in the present. Therefore, most of the surrounding area was defined as a buffer zone to preserve the open spaces between the remains of Crusader structures for visual and ecological reasons, recognizing the land's critical importance as a nature reserve. The remains of the primary built-up sites and their adjacent areas were defined as separate organic cultural landscape polygons, each surrounded by a buffer zone.

Mi'ilya—the village of Mi'ilya was built around a Crusader castle originally constructed at the beginning of the twelfth century. The king owned the castle, hence the name *Castellum Regis* (in French: *Château Du Roi*). In the second half of the twelfth century, the castle was the center of a large independent estate, which included 36 villages in the western and central Galilee. Shortly before 1280, the property passed into the hands of Joscelin de Courtenay III after he married Agnes of Milly, heiress to the castle. De Courtenay bought several adjacent properties (including Hunin, Manot and Deir al-Assad) and was the most prominent landholder in the Crusader Kingdom of Jerusalem. The castle passed into the possession of de Courtenay's brother-in-law, Otto von Schoenberg, and he sold it to the Teutonic Knights. It was during this period that Montfort Castle became an important site. In addition to its role in the feudal system, it was a large fortress overlooking a secondary road that connected Acre and Damascus. The castle had a square typology, and three corner towers were well preserved. The northwest tower maintains a height of three

Fig. 5.44 Montfort Castle

stories with embrasures. The castle remains are on a *tel* in the center of the village, and other parts of the ruin were incorporated into a number of the village buildings presently in use. The village mosque, built during the Mamluk period, was constructed with Crusader-era stones.

Massive building on and around the remains of the castle has compromised our ability to see the Crusader-era history. Although the site is protected as a declared archaeological site, the protection of the landscape is only enforced as long as the archaeological findings are buried. Limiting new construction on and around the castle remains would be appropriate. If construction is necessary for reasons approved by the planning authorities, consideration should be given to the visibility of the castle across the western Galilee.

Montfort Castle (Qala'at Corian)—the remains of a Crusader castle can be seen in the heart of the spectacular Kziv scenery, sitting atop a steep spur between Nahal Kziv in the north and a deep gorge in the south. Established in the middle of the twelfth century as a fortified farmhouse that belonged to the King's Castle in Mi'ilya, its importance increased over the years, acquiring the name "the King's new castle." Saladin conquered the site after his victory in the battle of Hattin (1187), though five years later, it was recaptured by the Crusaders. In 1229, the castle was sold to the Teutonic Order and named *Montfort*, "the strong mountain" in French. The Germans translated its name to *Starkenberg*. The Teutons transferred their headquarters from Acre to Starkenberg during the thirteenth century, along with their treasures and archives. With the help of contributions, the Teutons were able to build an elaborate castle, which ranked among the most beautiful in the Holy Land. Baybars, the Mamluk Sultan, failed in his first attempt to capture the site in 1266. He returned in 1271, and the knights surrendered after a week of fighting. According to the terms of surrender, Baybars gave the knights safe passage to Acre, along with their treasures and archives.

In the early twentieth century, the castle was partially excavated along with a winepress, two moats and a paved road leading towards Mi'ilya. A significant threat to the remains of Montfort Castle has been the dense overgrowth that hinders the archaeological conservation and the preservation of the cultural landscape. In 2004, after addressing the difficult dilemma between preserving nature or preserving cultural values, the Israel Nature and Parks Authority cut down many trees covering the fortress, greatly increasing the structure's visibility.

Cultural Anchors

The Agricultural Farm—at the foot of the castle, on the banks of the Kziv stream, is a well-preserved two-story structure. The Gothic-style arches and steps leading to the second floor have partially survived the ravages of time and are the remains of a farm building built for the fortress and the mill. Next to the buildings are the remains of a dam that created a pond that fed the canals that powered the flour mills and irrigated the fields. The remains of many flour mills are located along the stream, and at least three can be linked to the Crusader period.

Khirbet Nachat—this was the site of the Montfort quarry, though there is also a Byzantine ruin onsite. The area was used as a stone quarry to supply construction materials for the nearby Crusader castle, and the remains are impressive. In the middle of the site is a hewn cistern from the Byzantine period, a drinking water source for the Crusader stonecutters.

Yehiam Fortress (*G'din*)—this site features the remains of an Ottoman fortress that incorporated remains from the Crusader period, built atop an ancient structure from the Roman and Byzantine periods. The fortress was a fortified farmstead and the surviving buildings are large and impressive, similar to other Crusader estates in the western Galilee. Today the site lies within the boundaries of Kibbutz Yehiam. Some scholars believe the site was under the rule of the King's Castle in Mi'ilya, while others suspect it belonged to a feudal estate that extended across the Beit Hakerem Valley. In 1208, the fortress was given to the German Teutonic Order and was destroyed by the Mamluk Sultan Baybars in 1265. Dahar El-Omar, the Bedouin Sheikh who captured the Galilee, took control of the site in 1738. When the founders of Kibbutz Yehiam came to settle the area, they lived in the ruins of the fortress, defending themselves against Arab attackers. Most of the fortress buildings visible today were built during the period of Daher al-Omar.

Khirbet Manot—this site contains the remains of a building used by the sugar industry during the Crusader, Mamluk and Ottoman periods. These impressive remains are an integral part of the broader story of the Crusader economy in the Galilee.

5.1.3.9 Towns Along the Carmel Coast: A Relict Organically Evolved Landscape Including Caesarea (Figs. 5.45, 5.46, 5.47, 5.48, 5.49 and 5.50)

Narrative

The narrow Coastal Plain along the western slopes of the Carmel Mountains is a well-defined area with diverse habitats and geological conditions. These have shaped human settlement since prehistoric times. In ancient times, the eastern basin of the Mediterranean was the cradle of civilization, and important civilizations that developed in the region depended on the sea for contact and influence in the world. At other times the sea was an arena of conflict over the control of both land and sea. The Carmel coast area strikingly reflects the history of human settlement in relation to environmental changes. It can be viewed as an open-air museum for the history of maritime trade in general and the development of ports in particular.

Fig. 5.45 Port towns along the Carmel Coast—polygon

The Landscape

The most exceptional features are the *kurkar* ridges (aeolianite, a calcareous sandstone formed from fossil dunes that were submerged and calcified through percolation of calcium carbonate-rich waters). The ridges extend parallel to the shoreline. The western ridges form a coastline constructed of eroded layers, rocky tidal areas, numerous coves and tiny islands. The ecological niche that developed in this landscape has a vital role. Historically, the streams flowing down the mountain erupted in numerous springs, creating a marshy area between the *kurkar* ridges. This ecosystem has been uniquely preserved only in this region.

During the ice ages that characterized the end of the Pleistocene period, the Carmel coastal area underwent

Fig. 5.46 The port of Caesaria and the port towns along the Carmel Coast

dramatic changes. It is possible to link changes that occurred during the Holocene, approximately ten thousand years ago, to human settlement patterns. A series of late Stone Age sites, now located one hundred meters from the shore and ten meters below sea level, highlight the importance of the prehistoric settlement of the Carmel caves. During the Chalcolithic period, settlements began moving eastward. At the beginning of the Bronze Age, the shoreline stabilized in its current location.

Since the Bronze Age, cities and ports were established along the coast, and maritime commerce flourished. The three major ports of Atlit, Dor and Caesarea represent the main periods of the ancient world. Secondary sites, such as Tel Nami and Crocodilopolis, also shed light on the complicated symbiotic relationship between humankind and the sea throughout history. An example of this can be seen in the lagoon off Dor beach, where 24 sunken ships from different periods were discovered.

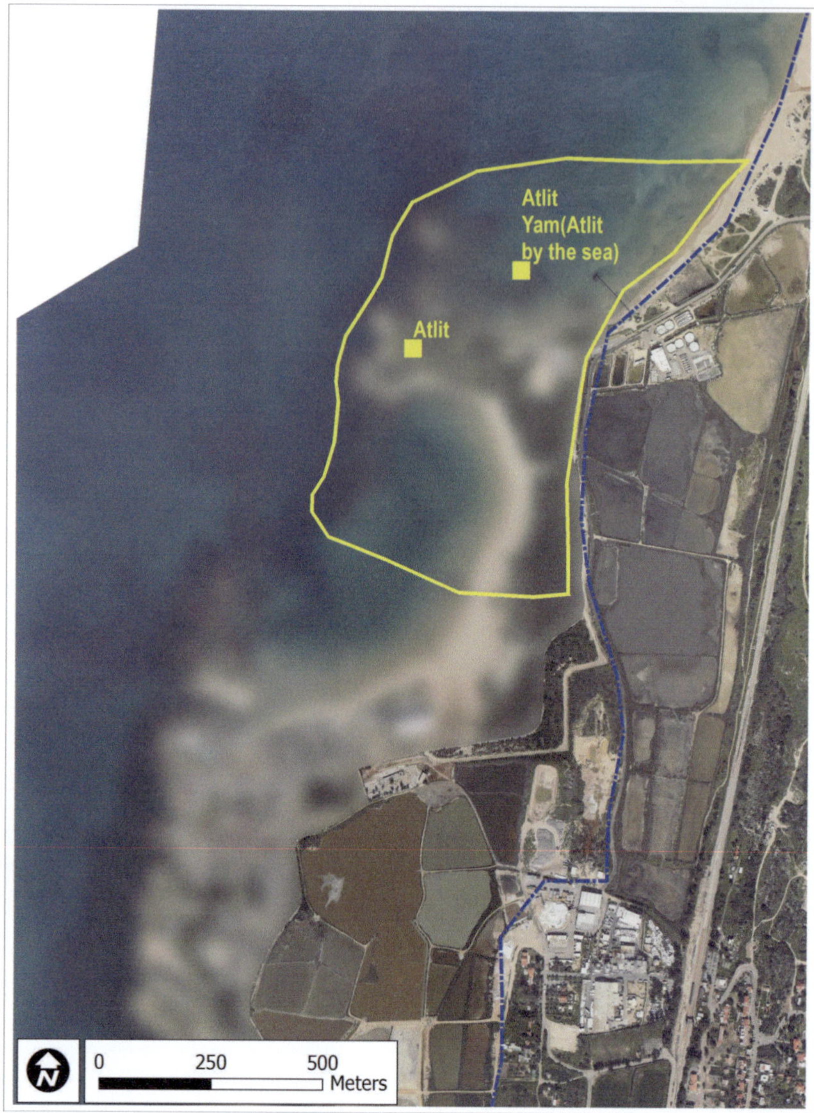

Fig. 5.47 Port towns along the Carmel Coast—Atlit

Defining Polygons

The polygon boundaries of the ancient port cities were generally defined according to archaeological finds on land that relate to a submerged seaport. The buffer zones run along the border between the central *kurkar* ridge and the eastern ridge. In areas where the *kurkar* ridge no longer exists, the boundary of the buffer zone follows the ridge line of the dunes along the shore.

While the Antiquities Law protects the submerged ancient ports and the Coastal Law largely protects the buffer areas, these two laws do not adequately protect the site from the intensive tourist development that has threatened to change the area's character completely. Kibbutz Neve Yam and other coastal communities are expanding and becoming privatized due to intensive development pressure along the coast. The town of Atlit was recently incorporated into the Hof HaCarmel Regional Council. Some see the salt ponds adjacent to the port of Atlit as a promising area for development. The military base near Atlit Yam underwent massive construction without being subject to the Coastal Law.

Fig. 5.48 Port towns along the Carmel Coast—Caesarea

There was no examination of the impact of construction on the surrounding cultural landscape.

Cultural Anchors

Atlit Yam—a submerged prehistoric village that represents a community of fishermen and farmers from the beginning of the Neolithic Revolution.

Atlit—an important Phoenician port that contained salt mines and later featured a Crusader fortress.

Dor—a large port built by the Takar seafarers in the twelfth century BC and flourished during the Persian, Hellenistic and early Roman periods.

Caesarea—an international port built by Herod and the country's capital during the Roman-Byzantine period. The city's port is one of the few ports of the Roman Empire that has been preserved, thoroughly researched, and excavated. Caesarea also contains impressive medieval remains.

Fig. 5.49 Port towns along the Carmel Coast—The Dor Port

5.1.3.10 Terraced Agriculture in the Judean Hills: A Relict Organically Evolved Landscape Including Ein Karem, a Village and its Cultural Landscape (Figs. 5.51, 5.52 and 5.53)

Narrative

The Judean Hills are part of the country's southern mountain ridge. The Israelite settlement period 3000 years ago and the construction of communities on the hilltops of the Judean Lowlands resulted in deforestation. The inhabitants diverted springs, built facilities to process agricultural produce and constructed stone terraces on the slopes along the base of the natural stepped rock.

There were four reasons to construct terraces. The first was to use excess stone from the site to create contiguous land units. The second was to moderate the slope's steepness by creating a sequence of stepped sections. Third, the terraces prevented soil erosion from heavy storm runoff. Finally, they slowed storm runoff to allow it to infiltrate back into the soil—excess water accumulated behind each step and infiltrated into the soil through the joints between the stones.

During the Byzantine period, which saw a dramatic increase in the country's population, many areas were covered with agricultural terraces that brought economic prosperity to the Judean Hills. Farms and agricultural facilities were built, even on previously marginal lands. Many

Fig. 5.50 Port towns along the Carmel Coast—Tel Dor

villages and agricultural areas have been continuously settled from the Chalcolithic period until the mid-twentieth century.

Following the establishment of the State of Israel, the widespread use of modern farming techniques led to the abandonment of many of the terraces. However, in the 1950s, large agricultural terraces were built in several places (near Tzur Hadassah and the Arazim Valley) as part of the afforestation efforts of the Jewish National Fund.

The Landscape

Alternating layers of hard and soft rock characterize the Jerusalem Hills. Their differential resistance to weathering created a stepped landscape. The soft rock layers formed expansive terrace treads, and narrow, hard dolomite or limestone formed the terrace risers. Water was channeled along the terraces where springs existed to irrigate the crops. The meager springs of the Judean Hills were only sufficient to irrigate limited areas. Where there was no irrigation, orchards were cultivated on the terraces featuring olives, figs, almonds and deciduous fruit trees.

The stepped landscape characteristic of the Judaean Hills was destroyed in many places due to afforestation projects, modern agricultural techniques, and construction.

Defining Polygons

The remains of agricultural terraces appear almost everywhere in the Judean Hills. Four areas were selected to represent the different ways the agricultural terraces have survived. Each one had a spring that irrigated the crops cultivated on the terraces beneath it, while olive groves were planted on terraces further away. A series of terraces in the Refaim Valley extend from the edge of Jerusalem to the village of Battir in the West Bank. In recent years the Jewish National Fund restored the slopes of Mount Eitan around the Sataf springs (Ein Bikura), planting olives and field crops.

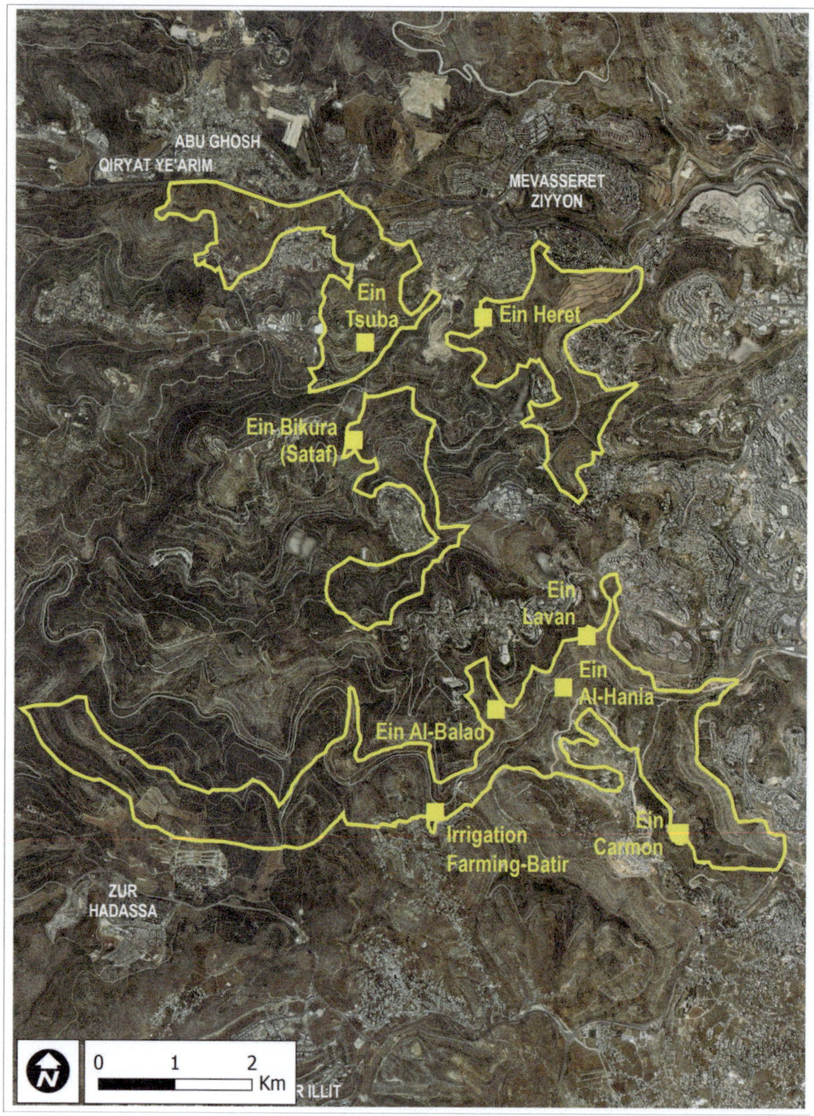

Fig. 5.51 Terraced agriculture in the Judean Hills—polygon

On Mount Harat the remains of terraces were partially covered by coniferous forest, though mature olives still exist and are cultivated on medium size plots. Around Tel Tzuba, there are remains of traditional terraces, though most now have deciduous trees and modern irrigation systems. It is also possible to see broad modern terraces created by heavy machinery that have damaged the ancient terraces but preserve the principles of mountain farming.

Cultural Anchors

Ein Bikura (Sataf), Ein Heret, Ein Tzuba, Ein Lavan, Ein Al Heenia, Ein Al Balad, Ein Carmon.

5.1.3.11 Liftah (Mey Naftoah)—Traditional Mountain Village—(Figs. 5.54 and 5.55)

Narrative

The village of Liftah, planted on the side of a steep mountain at the entrance to Jerusalem, is a historic complex built within the surrounding cultural and natural landscape. The village has unique cultural, historical, architectural and scenic values that reflect the development of rural settlement culture in Judea from ancient times to the twentieth century. Liftah is an example of a historic village frozen in time, since the middle of the twentieth century. Due to the

Fig. 5.52 Terraced agriculture in the Judean Hills—Battir

Fig. 5.53 Terraced agriculture in the Judean Hills—Sataf

Fig. 5.54 Liftah (Mey Naftoah)—Traditional mountain village—polygon

historical events of 1948 and its geographical location, Liftah was preserved with extraordinary integrity. Unlike other Palestinian villages, it was not deliberately destroyed or changed beyond recognition due to modernization processes.

Abandoned seventy years ago, the village has become a unique ecological site over time as spontaneous vegetation has grown over its ruins. Since no other villages remain in this state, it is important to preserve this unique place, especially in light of recent development proposals.

The Landscape

The village of Liftah is located in a rocky area at an altitude of 650–730 above sea level on a steep slope above Nahal Mey Naftoah. The stream valley drains into Nahal Beit Hanina from the southeast and connects to Nahal Sorek. The natural factors that enabled the settlement include a permanent and abundant water source, the presence of caves and rock shelters used for shelter and residence, the availability of stone for construction, a steep slope that allows protection and observation and proximity to fertile valley lands suitable for cultivation. The village was founded nearby due to its proximity to major roads: an east–west road leading from the Coastal Plain to Jerusalem (now Highway No. 1) and the road south to the Ella Valley (now Givat Shaul Street). Its proximity to the Jaffa-Jerusalem Road significantly affected its development and economy.

The main rock formations in the village areas consist of chalk and Hawari chalk (Kfar Shaul formation) that create a steep and rocky landscape, on which it is difficult to build, and is usually found in the upper elevation of the village slopes. Gray dolomite rocks (Aminadav formation) were used to construct the village buildings. They form wide

Fig. 5.55 A view of Liftah today

horizontal rock steps and a convenient substrate for paths and historic roads. The rock steps were also used for agricultural terraces. Liftah's geological characteristics provide a unique fabric of natural stairs, terraces and built fences, residential buildings and roads that together create its unique cultural landscape. The village structure preserved the natural drainage system and prevented soil erosion, thus protecting the natural topography of the valley and slopes.

The village location, just west of the national drainage divide, supported a variety of habitats in the transitional area between Mediterranean vegetation in the western Jerusalem mountains and desert vegetation. The diverse plant formations (Mediterranean woodlands, scrub, orchard areas and wet areas, such as Liftah Spring) have created rich wildlife habitats.

As part of the conservation survey of the village, Liftah's wildlife was surveyed and documented. A relatively diverse concentration of reptiles, rodents, birds, butterflies, and even substantial populations of large mammals such as jackals, foxes, porcupines and deer were found in the open spaces. Many unique and rare species of vegetation and wildlife were found.

Over the years, the remnants of agriculture and ancient settlement have created unique plant communities that combine ancient cultural species (such as fig, nettle, and almond) and native plants typical of the area. The habitat of the spring and its slopes is the most developed of all the springs of Jerusalem and includes an impressive concentration of brook willow and fig trees. The scrub areas (characterized by thorny burnet and wild thyme) are significant for the increasingly rare animals in the Jerusalem Hills area that have been impacted by the disturbance caused by development and new plantings. The plant community of almonds and nettles is common among the ruins of the village. One of the country's largest concentrations of this plant community is found at Liftah.

Defining Polygons

Liftah's fields extended down the Sorek stream valley (Nahal Sorek) and up the mountain in the direction of the present-day Romema and Kiryat Moshe neighborhoods. Its buildings gradually ascended towards the city and were built along the main traffic routes to the city. Today the village is bounded by major arteries that very clearly define the boundaries of its cultural landscape: Highway 1 and Begin Road to the south and east, and the railway and Road 9 to the north and west. This area preserves the unique variety of vegetation and the view of the terraces that characterize the village landscape.

5.2 Cultural Landscapes with a National and District Level of Significance

5.2.1 Designed Landscapes

5.2.1.1 Field of Giants (Sde Refaim) (Figs. 5.56, 5.57, 5.58, 5.59, 5.60, 5.61 and 5.62)

Narrative

Dolmens are large stone monuments that have been discovered in various sites around the world. The flat basalt plateaus of the eastern Upper Galilee (Ramat Korazim) and the Golan have yielded one of the largest concentrations of these monuments in the world. During the winter, the area is covered by a rich meadow and serves as a pasture.

In the ancient Breton language, "dolmen" means "table of stone." Near the ancient site of Gamla, "classic" table-shaped dolmens were discovered. They are constructed of two large upright stones with a horizontal stone placed on

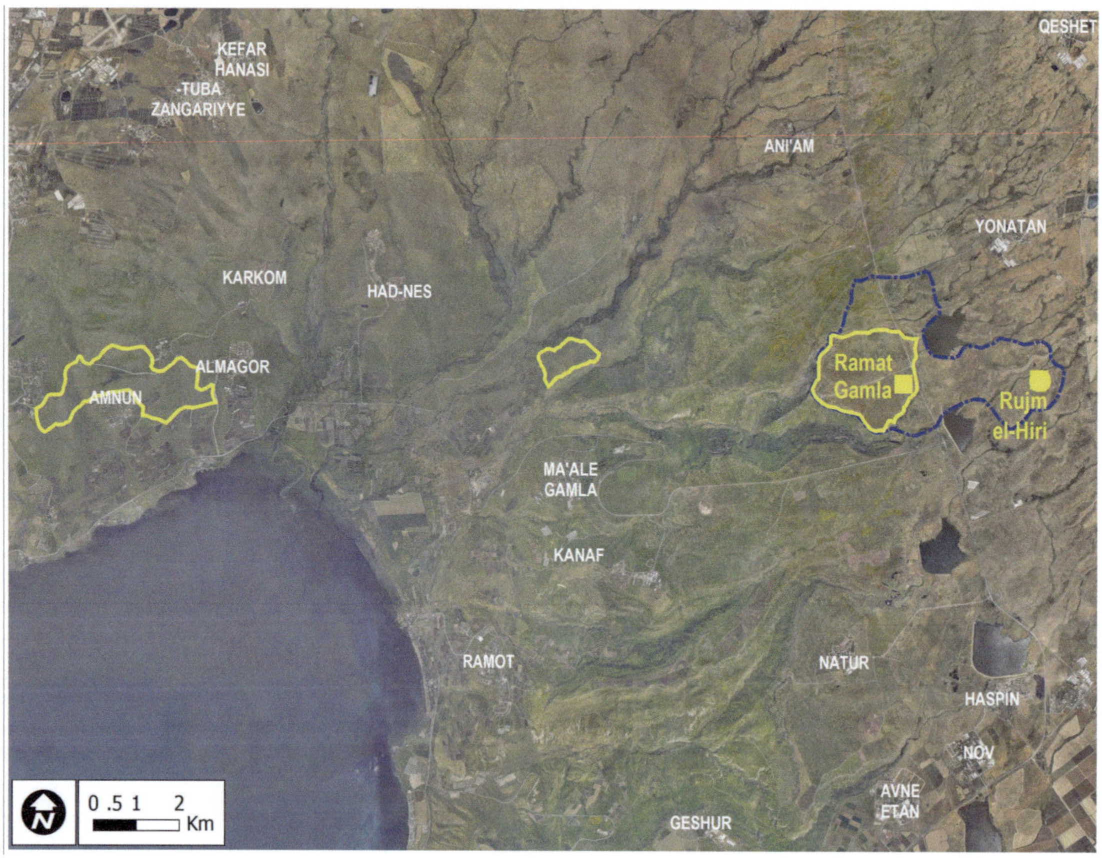

Fig. 5.56 Sde Refaim—polygons

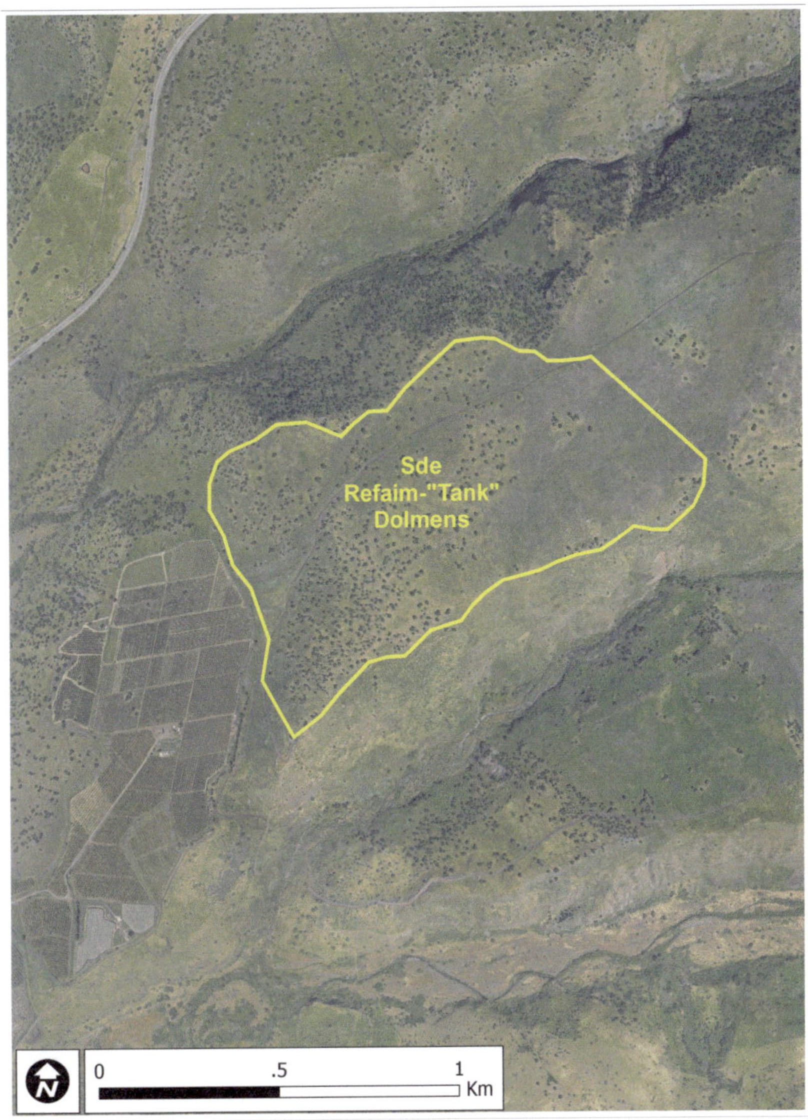

Fig. 5.57 Sde Refaim—tank dolmens—polygon

top. More complex dolmens may be composed of six or seven large stones. Occasionally the dolmen is sunken into the earth, appearing like a subterranean chamber with a narrow corridor leading into it. Some dolmens are entirely covered with piles of stone (*tumuli*). Archaeological excavations have established that the dolmens served as Bronze Age burial sites. Burial in the dolmens was a secondary burial after the dead were initially interred in the ground. Their bones were gathered later and placed in the dolmen, along with other objects that served the deceased throughout their lives. Archaeologist Claire Epstein attributed the structures to nomadic shepherds, claiming that dolmens are their primary imprint on the landscape (1974). Contrary to this view, archaeologist Lipaz Vinitzky believes that only city dwellers could have constructed such complex structures (1991).

An unusual type of dolmen was discovered on the slopes of the Batera stream in the Golan Heights. These are referred to as "tank dolmens" due to their shape. Their roofs appear as steps that jut out from both sides, creating a pediment-like form. These roof stones are quite large and covered with a tall pile of rocks. Inside each dolmen, a chamber was unearthed measuring up to 7.5 meters long and up to 1.5 meters wide. At its highest point in the center, the chamber was taller than the height of a man.

Close to the large field of dolmens in Ramat Gamla is Gilgal Refaim ("Wheel of Giants," or *Rujm Al Hiri*), a site composed of three perfectly concentric circles with stone

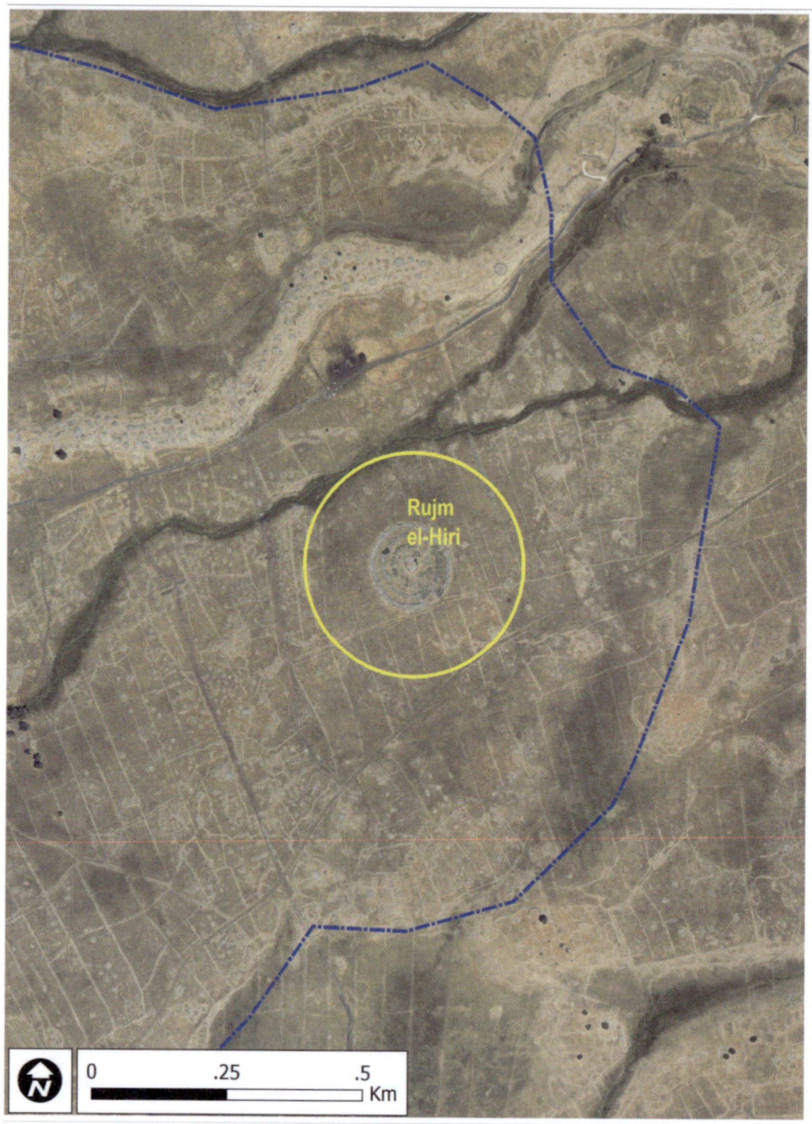

Fig. 5.58 Sde Refaim—Rujm el-Hiri

walls reaching a height of two or three meters. At the center lie the ruins of a circle, part of which is constructed with megalithic stones. The outer circle's circumference measures 156 meters and has two large openings. At the center of each opening is a large pillar of stones, while at the center of the site, another, larger pillar rises to a height of seven meters with a hidden burial chamber beneath it.

Gilgal Refaim is one of the great unsolved archaeological mysteries of the Golan. Archaeologists hypothesized that the site, like the dolmens surrounding it, constructed during the Bronze Age. Some believe that a significant personage from that period was buried here, while others believe the site was an ancient pagan stargazing site. The source of the name given to the site, Gilgal Refaim, is found in the biblical book of Deuteronomy (Chap. 2: 11), "these also are accounted Rephaim, as the Anakim;" the name of the whole region, Sde Refaim, is derived from the biblical passage as well.

Defining Polygons

The dolmens are distributed over an extensive area, and the Antiquities Law protects every artifact. The first cultural landscape polygon was drawn around Ramat Gamla and the second around Ramat Korazim, the areas with the highest concentration of dolmens in the region. The cultural landscape

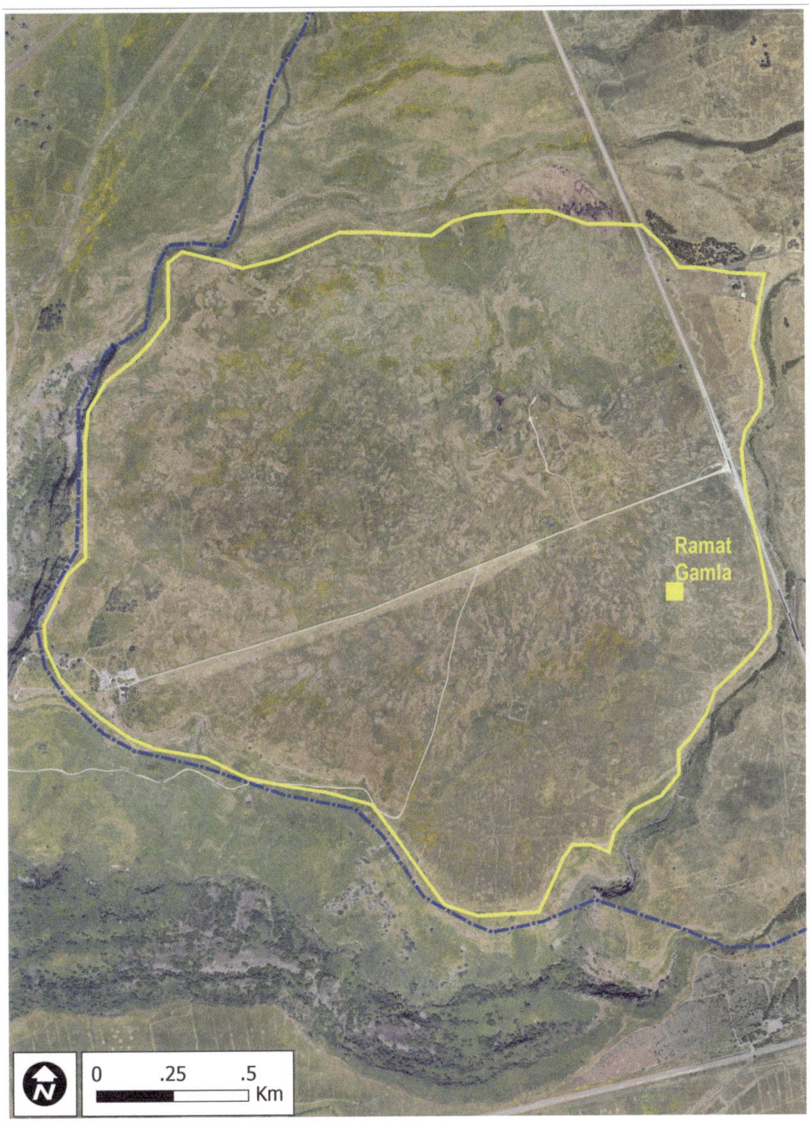

Fig. 5.59 Sde Refaim—Ramat Gamla—polygon

polygon defined around Gilgal Refaim is a circle with a radius of 200 meters and includes the ancient circle. The buffer area links the two sites creating a single cultural landscape that represents the remains of the monumental rituals of the Bronze Age that left their mark on the hills of the Golan.

Both the polygons and the buffer area lie within an area that serves as a military training zone. The primary threat to the existence of the artifacts in their landscape context is their visible scarring by military maneuvers, which affects the area's attractiveness. The site cannot currently be opened to the public as a visitors' site, and it is slowly deteriorating.

The Ramat Korazim dolmen polygon lies in an area zoned for agriculture. In the past, many dolmens were destroyed when the land was cleared to plant orchards. Additional land clearance may harm the willingness to protect the site as a cultural landscape. There is a conflict between agricultural uses and the protection of the site as a cultural landscape.

Cultural Anchors

Gilgal Refaim, the Monumental Dolmen in Ramat Gamla, Khirbet Berekh.

Fig. 5.60 Sde Refaim—Khirbet Berekh—polygon

5.2.1.2 The Sachne (Gan HaShlosha—Garden of the Three)[2] (Figs. 5.63, 5.64 and 5.65)

Narrative

The Sachne (Gan HaShlosha) is a national park in the Beit She'an Valley, located at the foot of Mount Gilboa, with an area of approximately fourteen hectares. The park's focal point is the Sachne Pool, fed by the waters of Amal Spring (Ein Al-Atzi), a spring sourced in a cave with an output of 1800 cubic meters per hour and a constant temperature of 28 °C. The pool's waters are clear and strikingly blue. They flow into the Amal stream, a few hundred meters to the east, passing through the heart of Kibbutz Nir David and then merging with the fishponds and irrigation system shared by communities in the region. The brackish water is not suitable for drinking or irrigation, but it is warm and pleasant for bathing during all seasons.

During the first half of the twentieth century, the Sachne was a small natural pool hidden among thorns, reeds, berries and a few jujube trees. The site was transformed by the influential Israeli landscape architects Lipa Yahalom and Dan Zur (Israel Prize laureates in 1998). Together with Zvi Bahir, one of the founders of Kibbutz Nir David, they turned the local natural landscape into one of the most beautifully designed cultural landscapes in Israel. Their concept was to design a landscape that would appear natural, featuring a

[2] This section was written by Nurit Lissovsky

Fig. 5.61 Sde Refaim

Fig. 5.62 A dolmen in Sde Refaim

Fig. 5.63 The Sachne—polygon

naturalistic, free style with rolling grass hills and shade trees. The ponds, paths, bridges, waterfalls and water crossings are made of concrete and stone, but their appearance, scale and composition appear natural. In fact, except for the water, which flows from the depths of the aquifer without any human intervention, every aspect of the the Sachne landscape has been planned and designed.

The river was widened and deepened, and three dams were built to create two additional bathing pools to the east of the original Sachne pool. The three pools are interconnected: near the spring, the upper pool is bounded on three sides by a rock cliff trickling with water. At the lower section of the pool, the water flows over a "floating" stone surface that, together with the waterfall it creates, has become the symbol of the park. The wide banks create an open space with shaded areas of lawn. The middle pool is stepped to create varying depths. The famous waterfall is located at the top section of the pool, and alongside it are wooded areas to provide shade and playgrounds. The bottom pool includes a mill, likely from the Ottoman period, which has been restored and continues to be powered by the water as in the past.

The Landscape

The establishment of the national park weaves together several parallel stories regarding recreation, archaeology, ideology and memory. Zvi Bahir was the man who conceived of the Sachne in its present form. In the early 1950s, Bahir recognized the stream's potential and decided to develop it for the benefit of the inhabitants of the valley. Bahir promoted the construction of the park, employing immigrants from the nearby development town of Beit She'an with funds from the Ministry of Absorption (including a master mason from Yemen who built the wall for the noted waterfall with stones from the ruins of the nearby

Fig. 5.64 The Sachne and Kibbutz Nir David

flour mills). Bahir also had carob trees planted to create a fruit orchard that would be eligible for funding from the Ministry of Agriculture.

Landscape architects Lipa Yahalom and Dan Zur brought Bahir's vision to life and gave it form. However, it is important to note the human interventions in the Sachne that preceded their work. In the days before the establishment of the State of Israel, the land in the region was owned by Arabs who established flour mills and were the first to expand the pool and create waterfalls to drive the millstones.

The park was inaugurated in 1958 as part of the celebrations for the first decade of the State of Israel, in the presence of the second Israeli President, Yitzhak Ben-Zvi. The park is called Gan HaShlosha (the "Garden of the Three") in memory of Haim Sturman, Aaron Atkin and David Mossenson, three local who were killed in 1938 when their vehicle hit a mine while driving to the nearby Kibbutz Tirat Zvi. Their names are memorialized on a large stone boulder near the spring.

Defining Polygons

The cultural landscape polygon encompasses the park compound designed in the 1950s. While the area is protected from any building that is not directly intended "for recreation in the outdoors or the perpetuation of values that have historical importance or archaeological, architectural or landscape values" (National Outline Plan 8), the architectural design of the park is not protected in any way, and could be redesigned. The history of Kibbutz Nir David is also an inseparable part of the story of the designed landscape of the

Fig. 5.65 The Sachne Pool

Sachne Pool as the kibbutz is adjacent to the National Park, and kibbutz members initiated its design.

The buffer area includes a tower and stockade site situated at the edge of the park that recalls the establishment of Kibbutz Tel Amal (1936), the first tower and stockade settlement erected in the valley during the bloody clashes between Jews and Arabs during the British mandatory period. The buffer zone also includes the Museum of Archaeology located on the ancient Tel Al-Asi (Tel-Amal). The museum displays artifacts excavated from the *tel* (and from other excavations conducted in the Beit She'an valley and a collection of Greek pottery from the Mediterranean Basin) and a narrow strip of fields and orchards between the park and the regional road. The buffer zone aims to prevent the landscaped park from becoming a green space trapped between built-up areas.

Cultural Anchors

The pools and waterfall, the restored mill, the Museum of Regional and Mediterranean Archaeology.

5.2.1.3 Avenues of Washingtonia Palms (Figs. 5.66, 5.67, 5.68, 5.69)

Narrative

The *Washingtonia* is a member of the palm family, named in honor of George Washington, the first president of the United States. The Israeli *Washingtonia filifera* originated in southern California and western Arizona. *Washingtonia* palms grow in disturbed habitats, such as abandoned orchards, on the edge of fields and along roadsides. The fruits are sweet, attracting animals that feed on them and disperse their seeds.

The Landscape

The *Washingtonia* is a tree particularly suited to avenues; the palms have a single tall, straight trunk rising fifteen meters with an impressive crown of palm fronds at the top of the canopy. Boulevards of *Washingtonia* palms were planted during the late nineteenth and early twentieth centuries to enhance communities established during the First Aliyah

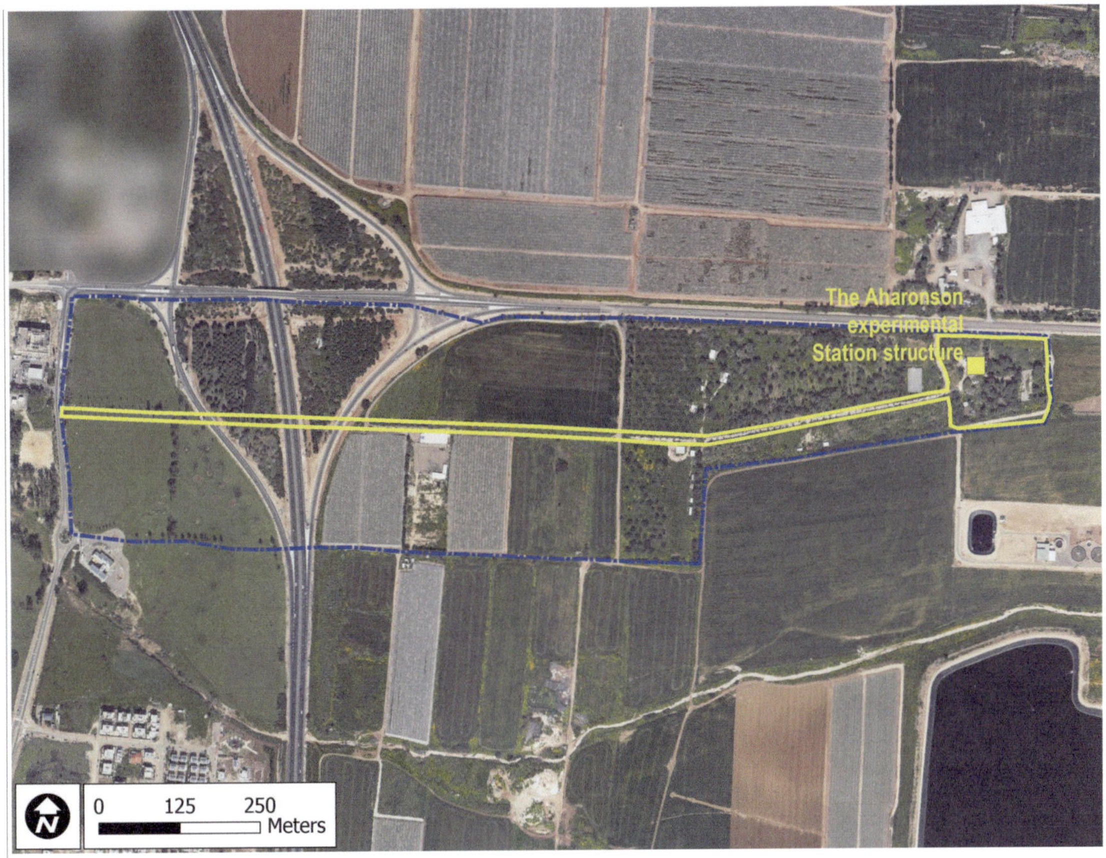

Fig. 5.66 Avenue of Washingtonia Palms, The Aaronsohn Experimental Station structure, Atlit—polygon

period. The avenues can be found at sites such as Old Hulda and the entrance to important sites such as the Mikve Israel agricultural school, the experimental station in Atlit, and German Templar settlements throughout the country.

Defining Polygons

Many avenues of *Washingtonia* palms were planted throughout the country. Some suffered neglect or were uprooted to make way for development. Three surviving examples were chosen as exemplars of *Washingtonia* palm avenues. The cultural landscape polygon that has been deemed worthy of preservation includes both the trees and the road. The older trees are at the end of their life cycle, and to preserve these avenues it will be necessary to plant new *Washingtonia* palms among the aging trees, and prevent planting other tree species.

For the avenue to have an impact on the landscape, it must be accessible and used by the public along its length. A buffer was defined around the avenues to ensure that no visible obstacles (including farm buildings, infrastructure facilities and antennas) are to be built that may harm the visual integrity and visibility of the avenues from surrounding areas.

"*Old*" *Hulda*—in 1905, the Jewish National Fund bought 200 ha of land from the Arab village of Khulda. Four years later, the Herzl House was built on the site, which served as a center to train young Jewish pioneers to work the land. A long and impressive avenue of *Washingtonia* palms was planted on the farm to delineate the route between the train station, Kfar Na'ana and Mazkeret Batya, which linked the farm to the modern world. The *Washingtonia* avenue is one kilometer long and surrounded by agricultural fields and vineyards.

Fig. 5.67 Avenue of Washingtonia palms, Atlit

The Agricultural Experimental Station, Atlit—the experimental station was established in 1910 under the direction of agronomist Aaron Aaronsohn, who lived onsite in a simple shed. A great avenue of *Washingtonia* palms leads west from the farm across the coastal road to the Khirbet Karta base opposite the Atlit Detention Camp. Although Highway 2 and the Atlit interchange interrupt the avenue, the kilometer-long avenue continues to be a significant landmark within the Carmel coastal landscape.

Bet She'an Valley near Kibbutz Sde Eliyahu—during the late 1920s, a group of German Templars from Wilhelma (today known as Bnei Atarot) purchased land to farm in the Beit She'an Valley. Settlers planted a row of *Washingtonia* palms along the road that led to their community, as was customary in other settlements. Yosef Weitz of the Jewish National Fund visited the site in 1932 and noted that five Templar families cultivated the land. After their unsuccessful attempt to grow bananas, they cultivated citrus and other fruit trees. *Washingtonia* palms are prominent in the landscape of the Beit She'an Valley, along with many date palm groves.

Cultural Anchors

Aaronsohn's Experimental Station building, the Herzl House—Hulda Forest.

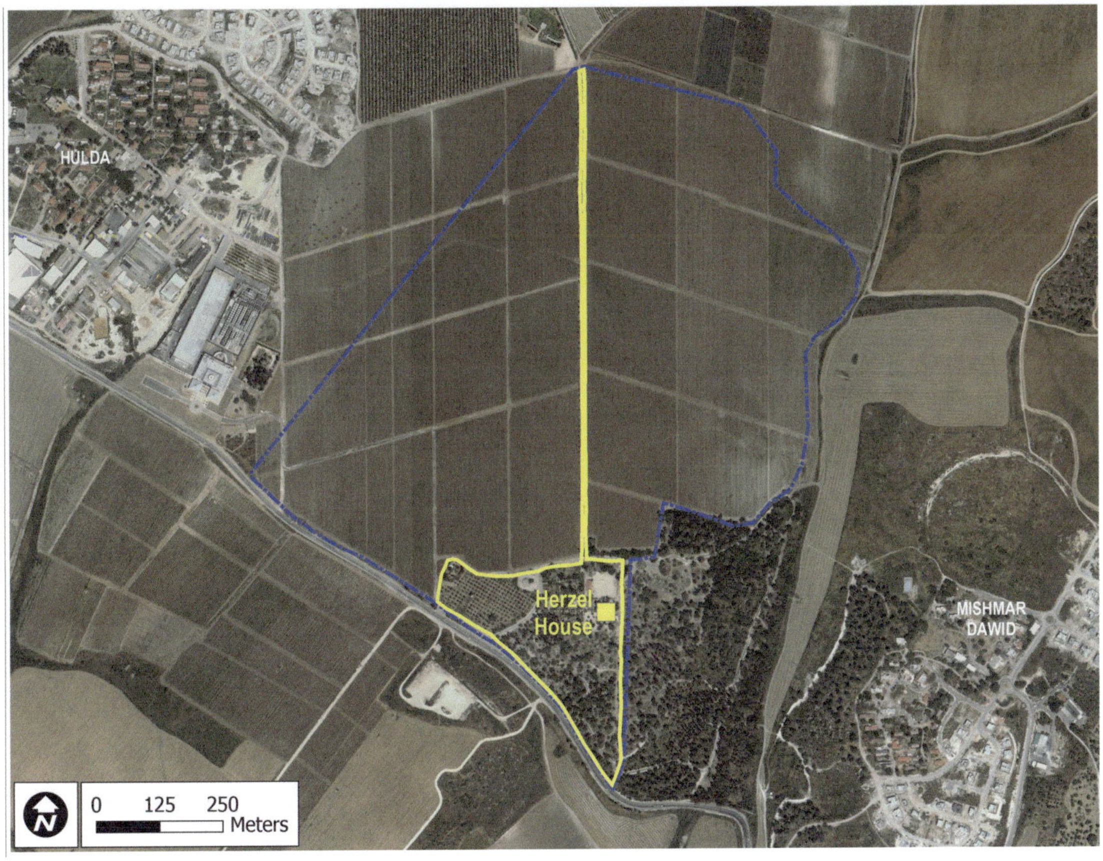

Fig. 5.68 Avenue of Washingtonia palms, Herzl House—polygon

5.2.1.4 HaNadiv Heights (Ramat HaNadiv) (Figs. 5.70, 5.71, 5.72)

Narrative

The memorial gardens for Baron Avraham (Edmond) Benjamin de Rothschild and his wife Ada (Adelaide) are located in Ramat HaNadiv, between Zichron Ya'akov and Binyamina. The location was chosen for its beauty and as a focal point for the Baron's activities in Palestine. The site is in an open area with natural and planted forests, historical and archaeological remains, and agricultural landscapes dotted with communities. The garden covers an area of approximately seven hectares and is among the most beautiful and well-kept gardens in Israel.

The history of the garden reflects the Baron's character (the "famous benefactor") and contains landscape architectural characteristics of the period. In 1914, during his fourth visit to the country, Baron de Rothschild indicated that he wished to be buried in the "rock" of the Carmel Mountains. In 1939, fifteen of the country's best architects were invited to submit design proposals for the burial site according to these guidelines. The plan had to include a tomb hewn into the rock, surrounded by four hectares of flowering gardens; the style and character of the structure were to follow Jewish tradition; the plan as a whole should express the position of Baron as a patron of the Yishuv in the Land of Israel; construction materials were to be local, and the plan needed to display the names and descriptions of the flora in the garden.

Architect Uriel (Otto) Schiller won the competition and invited landscape architect Shlomo Oren (Weinberg) to collaborate in designing the garden. The establishment of the garden was halted due to World War II, and it was only completed in 1954. Then, twenty years after their death, Baron de Rothschild and his wife were laid to rest in Ramat HaNadiv.

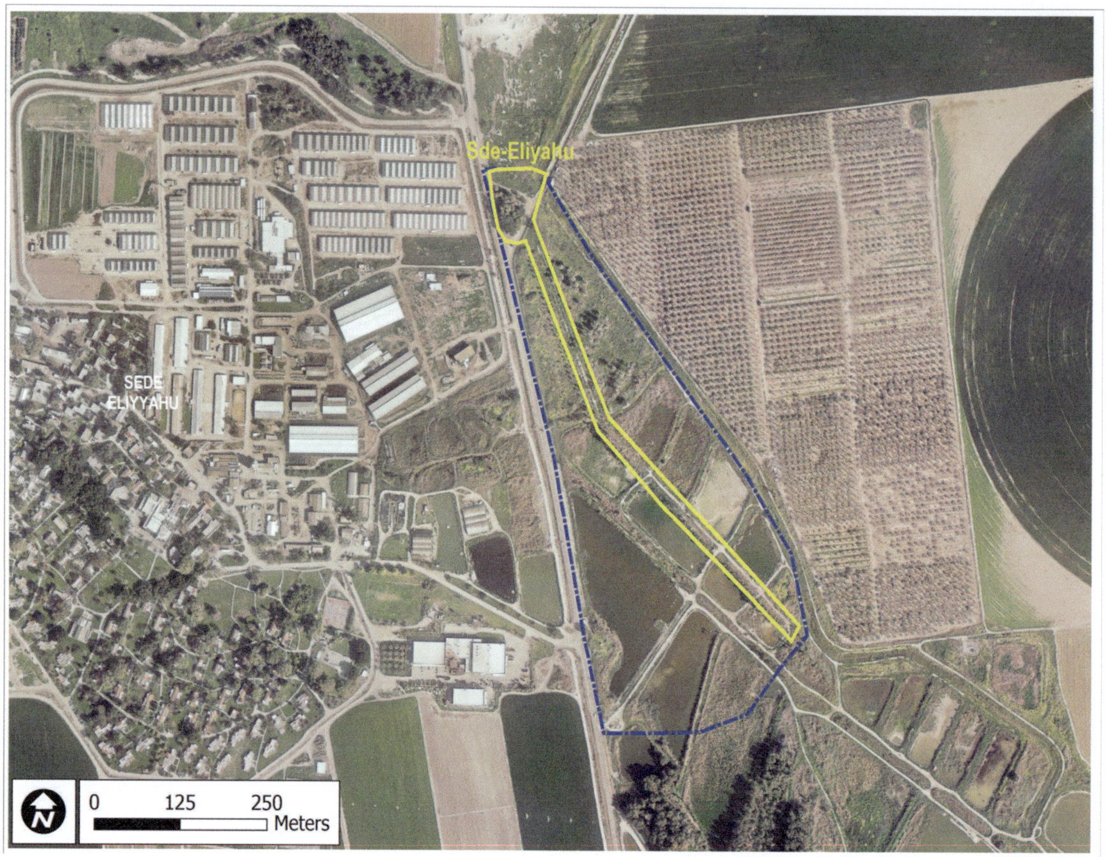

Fig. 5.69 Avenue of Washingtonia palms, Sde Eliyahu—polygon

The Landscape

The garden includes a variety of spatial experiences in a series of themed gardens, each with a different design. An ornate iron gate greets visitors, featuring the Rothschild coat of arms, a bronze shield supported at both ends by a lion and a unicorn with a crown of apples above. The coat of arms is divided into quarters that contain reliefs of an eagle, a lion and a clenched fist holding five arrows. Under the shield are the Latin words *concordia* (unity), *integritas* (integrity) and *industria* (diligence), the guiding principles of the family. The entrance plaza is paved with stone pavers divided into five longitudinal strips separated by narrow grass joints. These bands symbolize the five Rothschild brothers and lead to a wide staircase with two circular paths that link the other gardens. The perimeter path connects the Waterfall Garden, the Rose Garden, the Palm Garden" and the amphitheater. Between the two paths are wide grassy expanses. One moves through a sequence of open and closed spaces that provide a variety of perspectives. Landscape gardener Shlomo Oren-Weinberg, who designed the garden, emphasized the contrast between the enclosed spaces in the garden and the openness of the surrounding landscape.

In an article entitled "The Art of the Garden: What is the Impression of Space?" Oren-Weinberg claimed that establishing a balanced relationship between open grassy expanses and spaces enclosed by trees and shrubs creates a variety of aesthetic experiences that transform the garden from a collection of plants into a work of art. To this end, he suggested, "including the … surrounding landscape by opening up views from the garden to the landscape beyond." The use of "borrowed scenery" is taken from the English and Japanese garden traditions, and is a central concept in the

Fig. 5.70 Ramat HaNadiv

Memorial Garden. In the Waterfall Garden, for example, the vista of the the Coastal Plain and the Mediterranean is integral to the experience of the garden.

Views to the open landscape beyond provide a clear contrast to the enclosure and intimacy of the tomb in the center of the garden. A narrow, shaded path carved into the rock leads to an inner courtyard that acts as an entryway to the tomb. A narrow ornamental pool filled with goldfish surrounds the rectangular courtyard. Columns surrounding the pool are made of dark basalt, while the paving stones are pale limestone. The entrance to the tomb on the edge of the courtyard is flanked by cypress trees reminiscent of a memorial flame. The tomb structure, designed by architect Uriel Schiller, is simple and impressive. Heavy stone doors (evocative of Second Temple-era tomb doors) lead into a dark tunnel that winds down to the gravesite. In a cave niche facing Jerusalem stands a simple tombstone of polished black basalt. The tombstone is a seven-sided polygon with the names of the Baron and his wife inscribed in white letters.

The Rose Garden, designed with precise symmetry in the spirit of a French garden, is located in the northeast corner of the Memorial Garden. Between the rose beds are five small round pools with fountains symbolizing the five brothers, and to the north is a larger pool and fountain, symbolizing the patriarch. At the edge of the garden is a statue of a

Fig. 5.71 The gardens at Ramat HaNadiv

woman bent over a sundial, and a curving stone pergola runs along the perimeter. Other themed gardens include the Palm Garden, with a variety of palm trees, including the *Washingtonia*. Prominently featured along their main streets, the *Washingtonia* became the symbol of the Baron's colonies. The Scented Garden, a later addition established in 1985, is not part of the original plan. It contains raised beds of pungent herbs and other fragrances and was designed for visually impaired visitors. The grass amphitheater features curved stone seat walls.

The memorial garden design is based on a clear distinction between the garden and the natural surroundings outside the park. This duality is continuously reaffirmed in the design principles: the contrast between large and small spaces, the opening and closing of perspectives to the landscape, the alternation of shaded and sun-filled areas, and the contrast of naturalistic and geometric styles. Weinberg-Oren was influenced by the Italian, French, and English gardens and adapted them to the Israeli landscape in combination with local influences. The general concept and unique design details, such as the combination of paving stones and grass joints at the entrance plaza, the dressed stone walls and natural rocks scattered among flower beds in the rockery, were considered a landmark work of garden design in Israel.

Defining Polygons

The definition of the memorial garden polygon boundaries was straightforward as a garden wall defines the area. Surrounding the garden is a buffer zone whose definition was

Fig. 5.72 The Rosthschild tomb, Ramat HaNadiv

similarly clear. The steep cliffs of Mount Carmel mark the western boundary. The cliff slopes gently to the east to the Binyamina-Zichron Ya'akov road.

Ramat HaNadiv is one of the first nature reserves in Israel. It contains diverse natural and constructed habitats: a tall cliff overlooking the sea, boulders and outcrops of tuff, woodlands, undergrowth and garigue, olive groves, almond trees, planted forests, open spaces and a well-maintained garden. The natural woodlands are dominated by Mediterranean species, such as oak, wild olive and carob trees, and shrubs, such as *Cistus* (rockrose) and *Calicotome* (spiny broom). A planted coniferous forest covers much of the site. The exposed tuff slopes are covered in scrub. The polygon and buffer areas are owned by a company established in 1958 known as "Ramat HaNadiv Gardens." The Ramat Hanadiv Act, legislated in 1958, dictates that the company "will maintain these lands and will manage them as a public garden and park dedicated permanently to perpetuating the memory of Benjamin de Rothschild."

While the law ensures the preservation of the open space for public use, the relationship of the garden to the natural surroundings on the Carmel is not protected. A buffer zone was planted with pine trees during the 1970s. The natural woodland was threatened when a major fire raged through the area in 1980. It was later developed as a park. The unique architectural design of the garden is not legally protected as the owners are permitted to redesign the garden.

Cultural Anchors

The Rothschild tomb.

5.2.1.5 The Ben-Gurion Tomb: A Designed Landscape[3] (Figs. 5.73, 5.74, 5.75 and 5.76)

Narrative

The memorial garden and gravesite of David and Paula Ben-Gurion are located on the edge of a steep cliff overlooking magnificent views of the Tzin Valley and the Avdat Plain. The garden is the material realization of the Zionist dream and reflects the conflict between conservation and development. It was designed by the landscape architects Lipa Yahalom and Dan Zur. The location of this 5-hectare

[3] This section was written by Nurit Lissovsky.

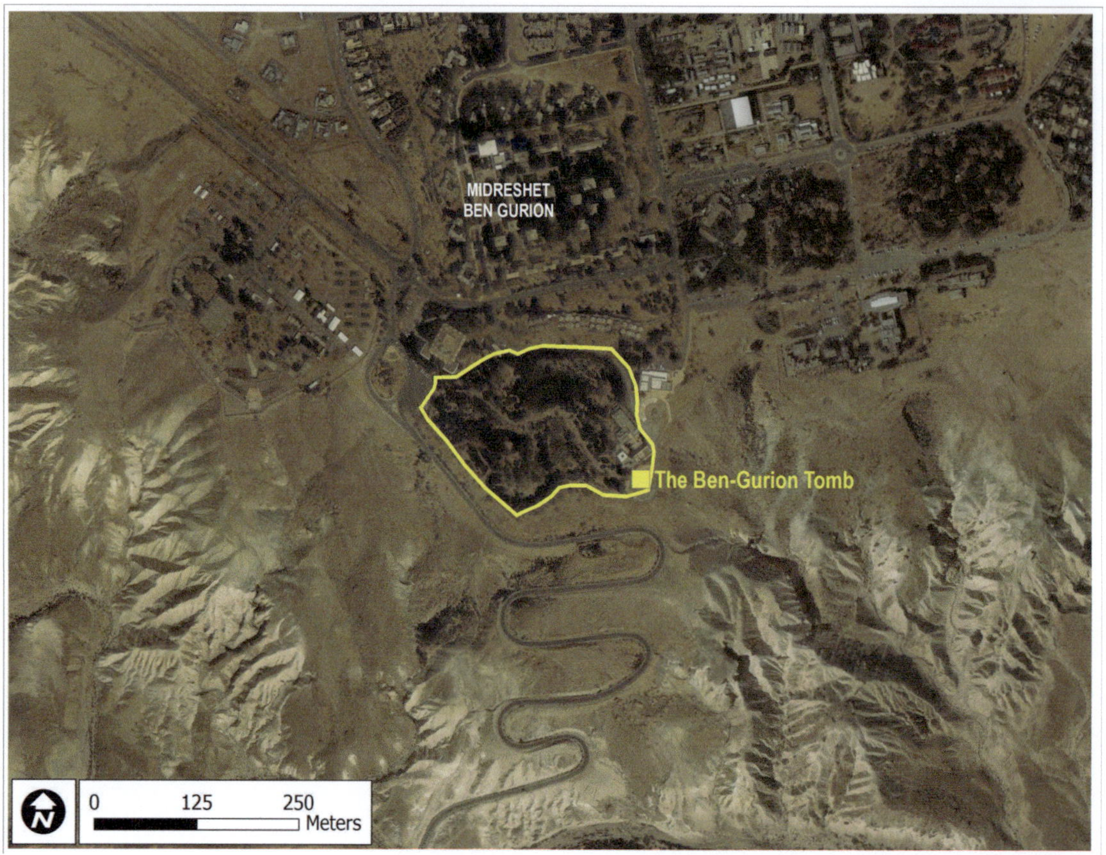

Fig. 5.73 The Ben-Gurion tomb—polygon

area was chosen by David Ben-Gurion, the first Prime Minister of Israel, who lived his final years with his wife Paula in the nearby Kibbutz Sde Boker.

The iconographic idea of the garden was to embody Ben-Gurion's vision, manifested in the concept of "making the desert bloom." The winner of a 1987 competition, the plan physically contrasts the barren desert with a vision of a garden, illustrating the tension between transforming the landscape and preserving its natural character. A dramatic sequence is created as one approaches the gravesite. From the secluded experience along the long, enclosed, winding path in the wadi, some 300 meters long, the view opens up at the gravesite, providing a broad vista over the vast Tzin desert. The design is executed with modest means. This simplicity is expressed by the restrained design of the gravesite and the use of desert plants adapted to arid conditions in muted shades of gray-green.

The Landscape

The planting concept is similarly modest in its expression. Olive, acacia and cypress trees differentiate the garden from the surrounding desert, but without any sense of artificiality. There are no flowers, for example, or other ornamental plants. The main path is designed to give the visitor a sense of walking in a stream channel using subtle textures of asphalt and gravel. Large boulders are placed along the path's edges to create a transition to the planted area. The boulders are all local, and the wadi appears natural, as though it had always been there. Grassy areas and shade trees were planted for the comfort of visitors. However, the idea of "making the desert bloom" does not draw its inspiration from the image of an oasis. The garden was designed in contrast with the desert environment rather than attempting to appear spontaneous or natural. This intentional rejection of the natural form of an oasis reflects

Fig. 5.74 Kibbutz Sde Boker overlooking the Tzin Valley

Ben-Gurion's vision of radically transforming the landscape. It is not a formal or ornamental memorial garden (such as the Ramat HaNadiv Memorial Gardens). The garden expresses the simplicity and egalitarian ethos of David and Paula BenGurion and the Labor Movement.

The garden's topography is similarly artificial; forty thousand cubic meters of earth were moved to create the wadi. Cypress trees were added to exaggerate the depth of the perspective. The plaza in front of the gravesite, which Lipa Yahalom called a "fan opening towards the landscape," is used for memorial gatherings. It is designed sparely with simple gravestones surrounded by the delicate texture of river-washed pebbles and large rock surfaces. The trees, which were planted to provide shade, and the planters, added later, somewhat diminish the intensity of the desert.

Lipa Yahalom and Dan Zur were awarded the Israel Prize for Architecture in the jubilee year of Israel's independence. The judges wrote that their work is "a great example of design solutions with an Israeli character, inspired by the early settlement in the country, local topography, and exceptional knowledge of the use of vegetation in this particular site." The design of Ben-Gurion's gravesite demonstrates their remarkable ability to transform abstract ideas into the material form of an "Israeli" cultural landscape.

Fig. 5.75 The approach to the Ben-Gurion tomb

Fig. 5.76 The gravestones of David and Paula Ben-Gurion

Fig. 5.77 *Liman* irrigation systems in the Negev—Be'er Chayal polygon

Defining Polygons

The cultural landscape polygon includes the landscaped park area surrounding the graves, and it is included in the Ein Avdat National Park. Both areas have statutory protection, but the design is not protected in any way, and the park managers can redesign it.

Cultural Anchors

The gravestones of David and Paula Ben-Gurion.

5.2.1.6 *Liman* Irrigation Systems in the Negev (Figs. 5.77, 5.78, 5.79 and 5.80)

Narrative

Limans are sunken basins created within desert river channels by constructing earthen dams to hold floodwaters. These basins create microclimates in which small groves of trees can flourish. The source of the word *liman* is the Greek word for a port, "a shallow pool of water," or figuratively—"a place to rest." Indeed, *limans* are like small oases, places to rest in the desert. The *limans*, a continuation of an ancient rainwater collection technique, were created as part of the Jewish National Fund afforestation program. They were established in areas where the annual rainfall is between 50 and 250 mm of rain a year. Trees are grown in a *liman* based "runoff harvesting." Because the amount of rainfall in desert areas is insufficient for growing trees, rainwater flowing downslope is collected and diverted to the *liman*. It is possible to artificially increase the amount of water in the soil to sustain tree growth.

The Landscape

The Jewish National Fund has constructed more than 250 *limans* in the Negev, creating green spots in the desert.

Fig. 5.78 *Liman* irrigation systems in the Negev—HaRo'a picnic area polygon

A typical *liman* is generally between .05 and 0.1 ha in size. It is planted with various tree species, including acacia, eucalyptus, tamarisk, palms, orchard trees, and several non-native species acclimatized in the *liman*. When heavy rain falls in the desert, the *limans* turn into small lakes until the water seeps into the ground and disappears. The *limans* are usually planted on the sides of roads and highways for use by travelers as rest stops. Bedouin shepherds take advantage of them as a place for their herds to rest and enrich their diet by grazing on wild plants.

Defining Polygons

Of the hundreds of *limans* scattered throughout the Negev, we chose the three most scenic examples: The Shepherd's Lot near Sde Boker, the *liman* in Ramat HaBoker, and the *liman* near the Bedouin town of Be'er Chil. All three are located in the central Negev, although this form also exists in the northern Negev and the Arava. The trees in the *liman* have grown large, becoming icons in the landscape and a regular rest stop for travelers, soldiers and shepherds. A significant number of the *limans* are located on the sides

Fig. 5.79 *Liman* irrigation systems in the Negev—Ramat Boker polygon

of roads that cross the Negev, some even within the road's right-of-way. The preservation of the *limans* is an important testimony to the constant struggle between man and the desert.

5.2.2 Organically Evolved Landscapes: Relict

5.2.2.1 Circassian Villages in the Golan (Figs. 5.81, 5.82 and 5.83)

Narrative

The ethnic community of Circassians settled in the Golan in the nineteenth century. Following the war between Russia and the Ottoman Empire, many Circassians were forced to migrate from areas in the Caucasus conquered by the Russians. As Sunni Muslims, they were welcomed to Ottoman lands. A small group of Circassian refugees moved from the Caucasus to Bulgaria. In 1878 they arrived at the port of Acre, hungry and destitute after an arduous journey. The Ottoman government settled them on land in the western Galilee and the Golan Heights.

The Circassians were loyal to the Ottoman administration and fought against Bedouin tribes who rebelled against the empire. The newcomers built villages, cultivated fields and raised cattle, displacing the Bedouin. A survey of the Golan conducted by the German engineer Gottlieb Schumacher (1883–1886) documented the Circassian settlement in Queneitra and approximately a dozen villages in the surrounding area. According to Schumacher, the village of Surman was comprised of three large Circassian villages, which contained some 200 buildings (900 inhabitants). Near the two southernmost houses, there stood a large cistern and the remains of an ancient building.

Fig. 5.80 View of a *liman*

The most important Circassian communities in the Golan, aside from Quneitra, included Beer Ajam, Bariqa, G'oiizeh, Jawziah Khishniyyah, Mumsiyah, Mansura, Ayn Ziwan, Surman, and Ruhina. The Circassians abandoned the Golan Heights during the Six-Day War.

The Landscape

The Circassian communities in the Golan formed a network of villages around the town of Quneitra. They differed from the villages of the *fellahin* (farmers) and the Bedouin. The Circassians were skilled builders and introduced the use of red-tiled roofs to the area. The design of their mosque's minarets was also unique. In many cases, these settlers reused ancient stones found in the area to build their homes and whitewashed their walls.

The Circassian villagers commonly used unique wooden carts pulled by oxen. The Bedouin had not used such carts; this technological innovation influenced the cultural landscape by enabling the paving of local roads. Ruins of the basalt houses and remnants of orchards and poplar trees near small springs are the current remains from this settlement period.

Defining Polygons

The village of Surman (Tsorman) was chosen to represent the material culture of the Circassian villages. The cultural landscape polygon includes the remains of the settlement: buildings, cisterns and streets. The buffer zone includes the road system to neighboring communities, isolated buildings, fields and avenues. The village of Surman is located just

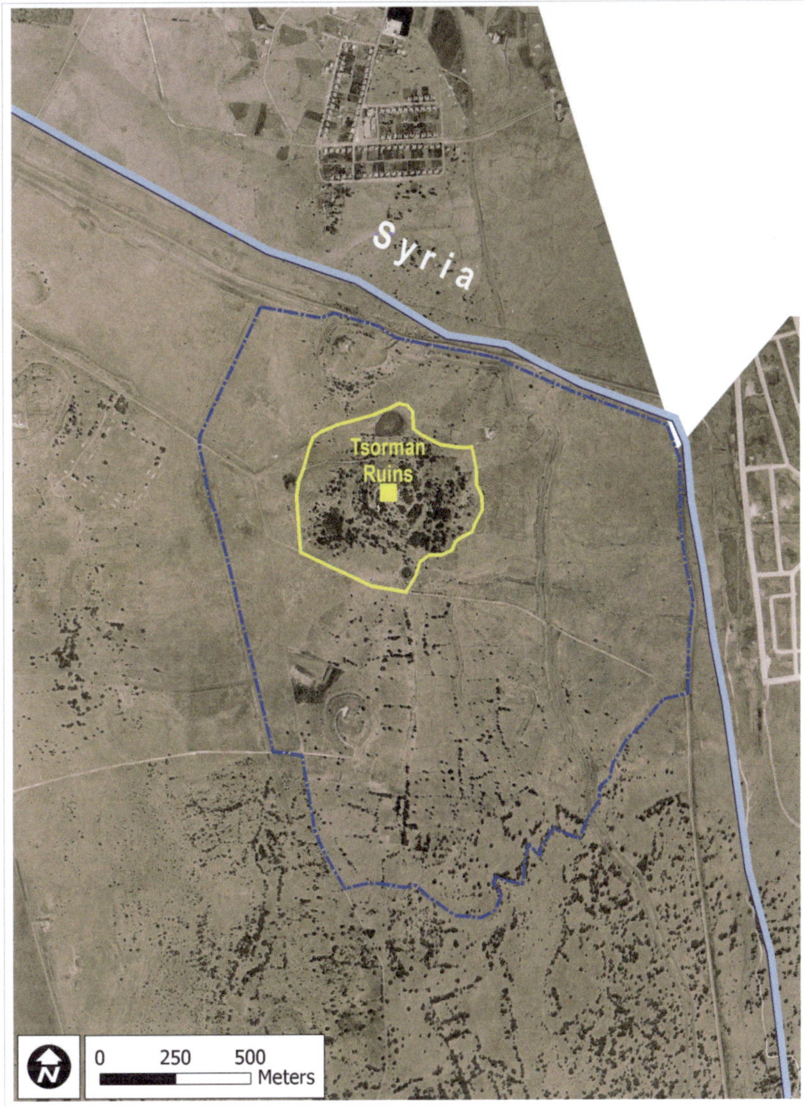

Fig. 5.81 Circassian villages in the Golan—Surman—polygon

north of the Bashanit Ridge, about two kilometers south of Quneitra near the ceasefire line, in an area where building is prohibited. The main threat to the existence of the cultural landscape is the neglect and deterioration of the remains due to the inability to turn the area into an organized site for visitors.

5.2.2.2 Agricultural Walls in the Central Golan (Figs. 5.84, 5.85, 5.86 and 5.87)

Narrative

For centuries, villagers living in the central Golan built stone walls around their cultivated plots, mainly vineyards. These were one meter high or more, built to enclose their plots and protect them from grazing sheep. There are also several watchtower buildings similar to the towers seen elsewhere in the country (*shomerot*). These are square elevated stone bases that were used to support temporary outdoor huts. During the summer, when the crops ripen, one could observe the area from these watch posts and protect it.

Defining Polygons

The agricultural walls extend over large areas in the central Golan, though most were destroyed due to modern farming on the land or military training. The best-preserved walls in the Golan extend between Mount Chazak (the Chazaka Ridge) in the east, the ruins of the village Ramatniya in the west, Mount Yosifon in the northern and Parza Hill to the south. This basalt plain has been fenced because of the possible presence of mines. Its enclosure accounts for the preservation of the walls in this location.

Between Highway 98 in the east and an antitank ditch in the west, there is a long narrow strip with a large concentration of walls. This strip was defined as a buffer zone. The

Fig. 5.82 The abandoned mosque in Surman

remains of the cultural landscape will be preserved as long as the walls remain in a military zone where no training takes place. As the area is not used for agriculture, it remains rich in natural vegetation and wildlife, including large mammals such as boars and wolves. In the center of the extensive area, defined as a polygon for conservation, there is a small area of some 60 ha, where the density and level of preservation remain high. In this region, without much effort, it is possible to revive traditional farming methods long practiced in the Golan and cultivate plots surrounded by stone walls.

5.2.2.3 Ancient Rural Settlement in the Golan (Figs. 5.88, 5.89 and 5.90)

Narrative

The Golan was extensively settled in the first century BC: roads were paved, villages were built, and land was cultivated for agriculture. Their remains are still apparent today. Several Jewish communities were also active in the Golan during that period that remained even after the Bar Kokhba Revolt. The Romans did not harm the villages that had not participated in the uprising.

At the very latest, the Jewish settlement in the southern Golan (the Hippos region) was abandoned during the third century CE. In the central Golan, the Jewish community reached its peak during the Byzantine period (the Talmudic period). Many remains of settlements and synagogues have been discovered dating from this period.

Following the Muslim conquest (seventh century CE), nomadic tribes gained control of the Golan Heights. The situation changed in the nineteenth century when Muslims from diverse ethnic groups began settling the Golan. These new immigrants usually settled on the ruins of previous settlements and incorporated ancient structures into their homes. Often, they repurposed the dressed stone from ancient buildings. During the Six Day War (1967), the residents abandoned their villages and fled to Syria, leaving behind village buildings built on ancient settlements.

The Landscape

Examples of Hauran architecture from the Roman and Byzantine periods can be found in the Golan Heights and further east. The most significant characteristic of this construction style is the use of stone roofs. Good examples of

Fig. 5.83 Detail of the mosque's minaret

these buildings in the Golan can be seen in Faraj, Ramatniya, Surman, Napach, Na'aran and Hispin. The best-preserved remains are found in the Kurdish village of Faraj. The community is an example of the unique hybrid architectural style found in the Golan Heights in which remains from the classic period are mixed with "modern" material culture.

Khirbet Faraj is located on the slopes of a small volcanic hill in the center of the Golan Heights, about two miles west of Tel Peres. The village and surrounding fields are visible from afar due to the distinct rows of trees and stone walls lining the fields and roads. The route leading from the village to the fields is also evident.

Among the ruins of the modern village, houses are scattered ancient buildings which have been well preserved. In some of the houses, two full stories survived along with the stone arches that supported their stone walls. It is also possible to see ancient burial caves and stairways carved into the basalt. Many crosses were found in the village alongside two lintels with lamps. It seems that in the Byzantine period, most of the villagers were Christian. However, there were also several Jewish families, perhaps because of its location on the road between the centers of Jewish settlements in the Golan.

Defining Polygons

The cultural landscape polygon includes the remains of the settlement itself: buildings, cisterns and streets. The buffer zone includes the road system to neighboring communities, isolated buildings, fields and avenues. The area of each village has been defined as an archaeological site. The pools of water have become an important example of nature conservation.

5.2.2.4 Grove Houses (Be'arot)—(Figs. 5.91, 5.92, 5.93 and 5.94)

Narrative

The citrus grove complexes reflect an important chapter in the history of settlement, agriculture and architecture in Israel. During the Ottoman period, there was widespread development in various agricultural sectors across the Coastal Plain, requiring constant attention and presence during the farming seasons. The first step was digging wells to irrigate fields, orchards, and citrus groves. There were often small structures built around the wells, and over time, additional structures were added to accommodate the needs of the farmers, such as warehouses and cattle barns. A residential building was added for the citrus grower, the workers and their families. These houses occasionally added a more elaborate second floor, with additional facilities and other buildings. A high stone wall surrounded all the buildings for security while creating a community structure for the farmers. The walls created built compounds that functioned independently in the agricultural landscape. These compounds were named *be'arot,* and later the entire grove was called a *be'ara* ("well" in Arabic) after the water source without which the grove could not exist. The original Arabic word was absorbed into the Hebrew spoken by the Jewish farmers. There were several *be'ara* initiatives by wealthy individuals or tyrannical government officials who acquired land or took over large tracts of land and held the workers as tenant farmers. At the center of these estates were mansions known as "palaces" and *khans*, sometimes reaching three stories.

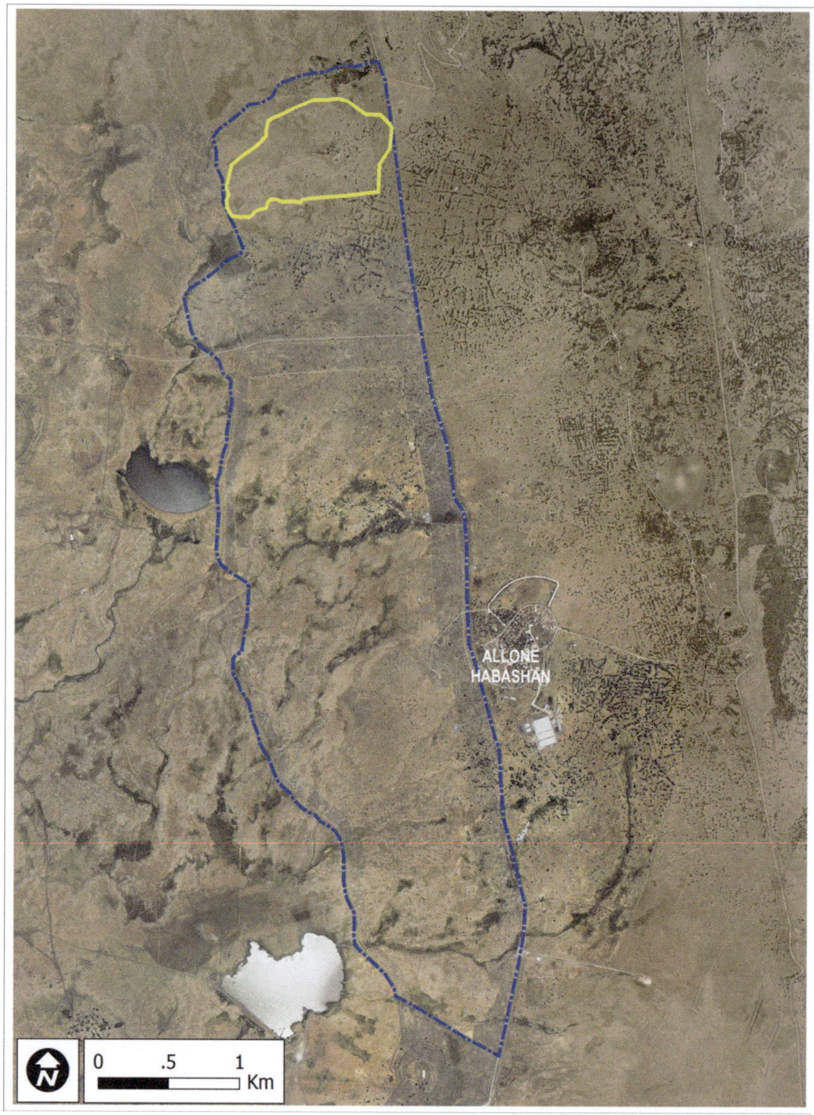

Fig. 5.84 Agricultural walls in the Central Golan—Alone HaBashan—polygon

The Landscape

"And many groves are surrounded by a high stone wall, and the eyes of the passerby only see the tops of the trees and the wheels of the wells and the buckets coming up, one after the other, pouring water into the pool in front of them. In most of the walls there are fixed water troughs with faucets connecting a pipeline to the pool. These are water fountains (*sabil*) that provide free water to thirsty passersby and each and every animal that follows them." (David Yellin, 1889).

At the height of the citrus industry boom in the heart of the Sharon region, hundreds of modest structures were built with local stone to create functional buildings. However, there were also dozens of larger structures and more complex *liwan*-style houses, some plastered and painted many times, designed with unique decorations typical of affluent homes in middle eastern cities. Most of the *be'ara* citrus grove compounds have been abandoned. Their remains are usually prominent in the landscape, particularly if the grove was uprooted in favor of field crops. In recent years they have become a haven for the homeless, while some houses were transformed into banquet halls. Despite the deteriorating physical conditions, some interesting elements such as ponds, fountains and wall decorations can be identified, and some have preserved unique building technologies.

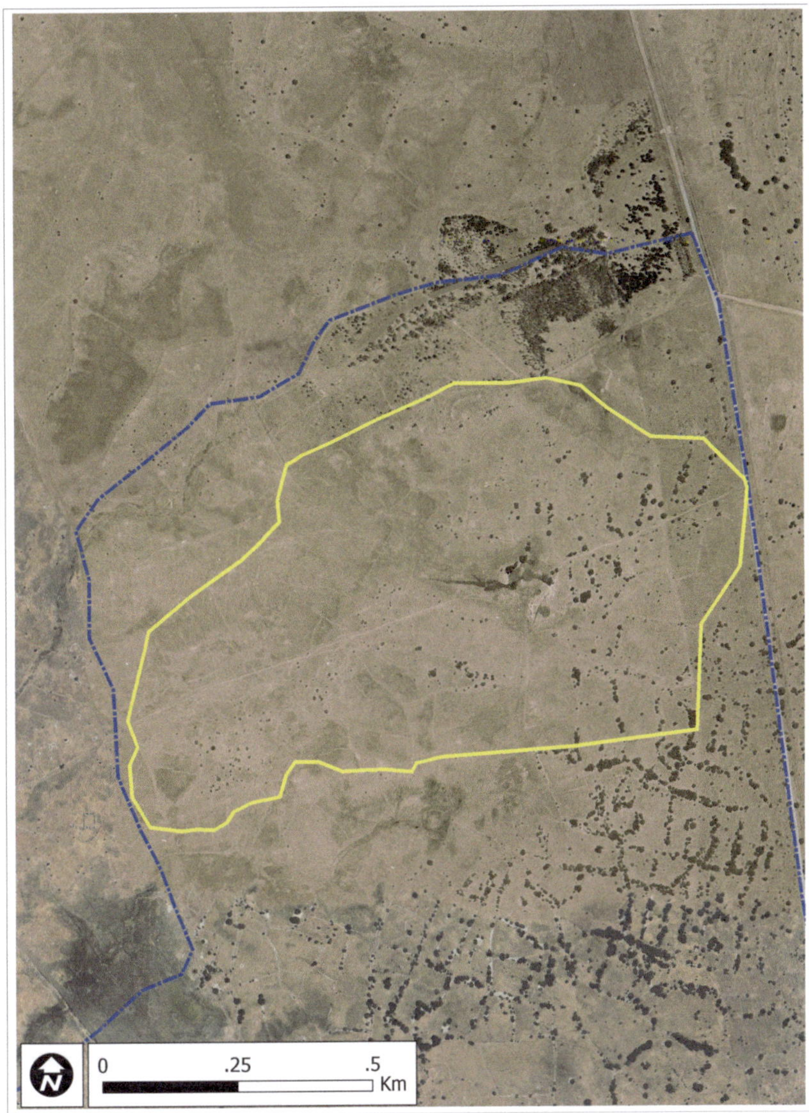

Fig. 5.85 Agricultural walls in the Central Golan—Alone Habashan—polygon detail

Defining Polygons

The *be'arot* are the last remnants of an Arab agricultural cultural landscape of citrus groves on the Coastal Plain. In recent years, private developers and architects have become more interested in restoring, preserving and managing these historic places for commercial and touristic activities. Out of dozens of grove houses whose remains exist along the Coastal Plain, three representative examples were selected. The choice was not based on a typological classification of the grove houses or clear criteria for preservation. There are dozens of other grove houses worthy of preservation. They should remain in the landscape by integrating them into the current planning and development of rural areas. However, even in cases where a survey was conducted, or professional conservation work was implemented, or the building was designated as an historic site, there is no formal requirement to maintain the landscape of groves that the *be'ara* served.

Cultural Anchors

Khan Zarkaniyya—the *khan* was likely built in the 1870s and is an excellent example of an agricultural settlement by Bosnian immigrants in Israel. In the early twentieth century, Jewish pioneers in the northern Sharon embarked on drainage operations. The Palestine Jewish Colonization Association acquired these areas to establish Jewish settlements. Groups of farmers settled the buildings on the estate's land

Fig. 5.86 Agricultural walls in the Central Golan

(Zaruniyya Farm) over the years, and their agricultural activities introduced various innovations. After the farmers left Zarkaniyya, it was turned into an experimental station directed by the Palestine Jewish Colonization Association to grow flowers for the perfume industry, grapes and fruit trees. The farm raised eucalyptus seedlings to stabilize the dunes of Caesarea, a project for which the Palestine Jewish Colonization Association was licensed by the British mandatory government. The site has been preserved exceptionally well and is currently undergoing a conservation process to convert it into a cultural center and residential building.

Burj Binyamina—Burj Binyamina was built in the late nineteenth century. In 1898, as part of his journey from Haifa to Jaffa, the German Emperor Wilhelm II and his entourage stopped there overnight. At the beginning of the twentieth century, Baron Edmond de Rothschild bought the Burj land from the heirs of Salim Khoury, a Lebanese-Christian. The lands were divided into twenty-seven plots for farmers from Zichron Ya'akov who lacked farmland. The farmers lived in the big *khan* in the Burj, whose remains still exist, and farmed the land. During the week, there was intensive agricultural activity at the Burj, though over the weekend, the farmers returned to the moshava some eight kilometers away. During the harvesting and threshing season, they stayed overnight until the beginning of the grape harvest, when they returned to the vineyards in the moshava. At age thirteen, the boys would join their fathers, sleeping and working together at the Burj, contributing to an atmosphere that molded the young, bold farmers. Today the ancient *khan* of the Burj is a banquet hall.

Bitan Aharon—the land originally belonged to Antoine Tian, a Christian from Jaffa, who owned expansive tracts of land in the Sharon region. In the heart of the area, he built the "Beit HaBek" or "Amarat Tian," which has survived

Fig. 5.87 Agricultural walls in the central Golan

until the present on the *kurkar* (sandstone) hill of the present-day Bitan Aharon. Joshua Hankin attempted to purchase the land at the end of the nineteenth century but succeeded only in the late 1920s. In 1930, the first settlement group settled in Beit HaBek, where they lived for about three years and then moved to establish the nearby Kfar Vitkin.

5.2.2.5 Grove House Estates (Figs. 5.95, 5.96 and 5.97)

Narrative

The potential of the red sands (red loam soil) in the Judean Lowlands and the Sharon region to support large-scale cultivation of groves was discovered in the 1920s. Four South African Jewish entrepreneurs sought to establish citrus grove estates. To this end, in 1928, they purchased land from Palestinian landowners from Kubibeh, Zarnuga, Sitria and Na'ana villages west of the road to Lod and the moshavot of Rehovot and Gedera. However, they were not familiar with the local conditions. They lacked knowledge of the citrus fruit industry, how to hire workers and manage the company. The South Africans required a foreman and workers, an agronomist and experts in citrus cultivation. Ephraim Eliash from Nahalal was recruited as a foreman, and in July 1928, the first small house was built for him and his family. A few months later, permanent quarters for workers were constructed. In November 1928, the first well was dug, and a celebration inaugurated the private agricultural enterprise.

Fig. 5.88 Ancient rural settlement in the Golan—polygon

The estate was called *Gan Raveh*, and the Yachin company prepared the land for cultivation, laying down an irrigation network composed of pools and canals typical of that period. They also established a nursery on the estate grounds and planted trees. While preparing the land for cultivation, they also laid the foundations for permanent structures. The first house built was called the Policinski House, a striking bright white, two-story building.

In 1935, when the grove began to yield fruit, the partners of Gan Raveh parted ways. Each partner received their share according to their investment in the partnership. Due to the outbreak of the Arab revolt followed by the Italo-Abyssinian War and World War II, the planting stopped, and the fruit could not be marketed. Like many other citrus growers, the investors of Gan Raveh tried to sell their property, and the grove changed hands several times.

The joint enterprise of Gan Raveh played an important role in transforming the entire region. Following the initial investment in Gan Raveh, additional private developers bought land and the Jewish National Fund invested in land purchases from the 1930s. Three working moshavim were established on these lands: Beit Hanan, Neta'aim and Beit Oved, which were part of the "Settlement of the Thousand" Plan to encourage settlement near the moshavot and the sites of private industry, chiefly for citrus production.

Several other institutions were established in the region, including the Einot Agricultural School and the Mordechai Farm Experimental Station, named after Mordechai Volcani, the brother of the head of the station, Isaac Elazari Vilkansky (Volcani), who headed the cultivation of field crops in Rehovot. The farm reflected the settling ideology of the national institutions. It sought to promote the systematic

Fig. 5.89 Ancient rural settlement in the Golan

study of agriculture, monitor and support the settlements, and adapt each agricultural industry to the local conditions of each region.

The Landscape

The surviving structures and agricultural facilities still visible in the landscape tell the story of the private enterprise that developed the lands west of the *kurkar* ridge in the late 1920s, as well as the development of Zionist institutions in the region in the years that followed. From the end of World War II until the 1990s, the "old road" between Gan Raveh junction and Yavne was dominated by the landscape of citrus groves. The Einot Agricultural School, originally established in 1932 to train women agricultural workers, also contributed to this cultural landscape.

It is still possible to identify the cultivation method used in the groves during the late 1920s and 1930s, known as the California processing method. It was characterized by widely-spaced planted rows that allowed a tractor to pass between them, irrigation channels, some of which were made from clay, and tracks for the *terenzinot*, carts laden with fruit that moved along sloping rails. The empty wagon was pushed uphill to be loaded with fruit and then rolled downhill to the packers. Many packing houses, pump houses and the manager's house were constructed from silicate bricks fired at high temperatures that replaced the carved limestone and sandstone.

Defining Polygons

The remnants of the groves and the intersections between private enterprise and communal settlements are scattered throughout the Coastal Plain. Based on the analysis of aerial photographs, maps, field surveys, documentation, written sources and interviews, an area was defined near the town of

Fig. 5.90 Ancient rural settlement in the Golan

Nes Ziona, where it is still possible to identify settlement processes from the late 1920s and the changes that occurred in the Coastal Plain following the rush of grove planting.

The groves west of the *kurkar* ridge made way for the extension of neighborhoods in the moshavot, the science and industry parks located in the city of Rishon LeTzion and the Regional Council of Gan Raveh. Much of the area between Highway 42 and Highway 4311 is still agricultural, and sections are still covered with groves. The small Coastal Iris Nature Reserve (*Iris atropurpurea*) was established to preserve the remains of an increasingly vanishing habitat on the *kurkar* ridge. However, the cultural landscape of the Grove House Estates currently lacks protection.

A buffer zone was defined around the heart of the agricultural area worthy of preservation, and three cultural landscape polygons were defined that also include the remains of a memorial.

Cultural Anchors

The Gan Raveh Estate—there are a number of landscape remains from the private enterprise at Gan Rave, including "The White House," which was repurposed as a banquet hall (HaAhuza). Three wells, three pools, and several packing houses remain from the groves around the farm.

Gurland Well—a well with a two-story structure above. The first floor held the pump motor, and the second story housed a residence. The building design is notable as elements of the international style are visible and correspond to its function. The well bears the name of Max (Mordechai Menachem) Gurland (1872–1940), one of three partners in the original Gan Raveh groves.

Mordechai Farm—a subsidiary farm founded in 1935 by Isaac Elazari Vilkansky (Volcani), the head of the experimental farm in Rehovot, at the top of the El Furan (HaTanur) hill, rising 28 meters above sea level. The farm included an

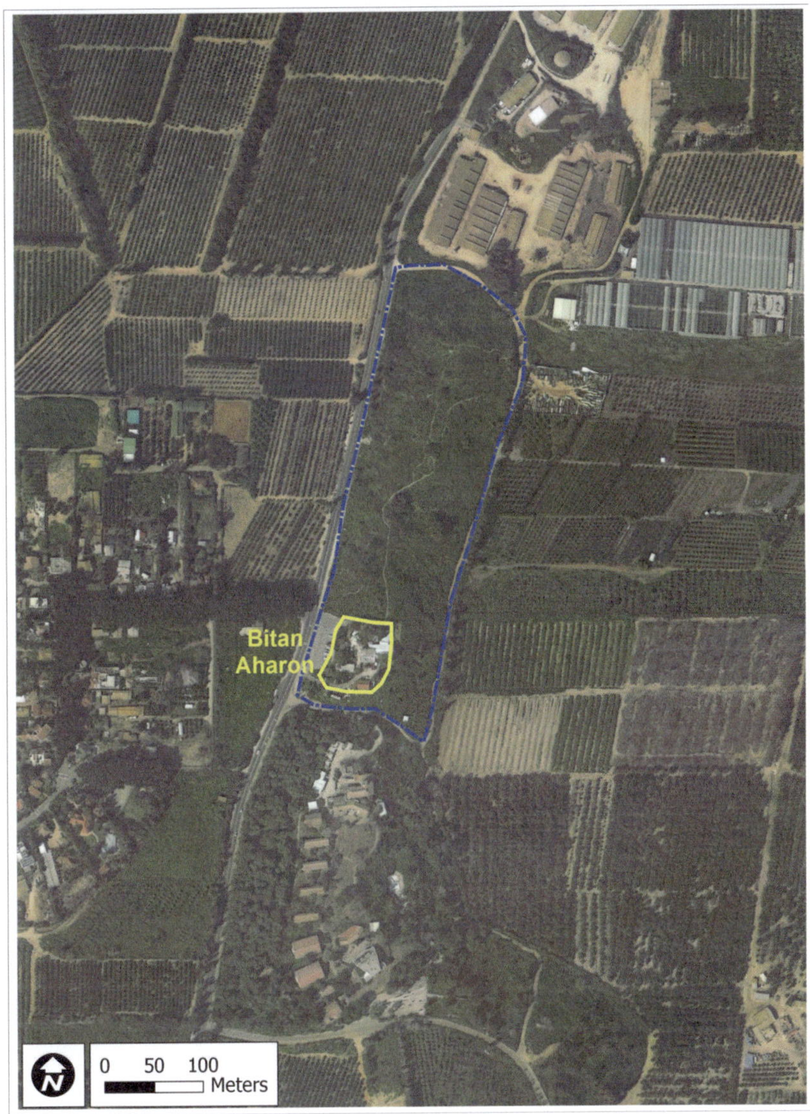

Fig. 5.91 Grove house—Bitan Aharon—polygon

educational training center for agriculture, animal husbandry, laboratories, pastures, crops and dwellings. The farm operated for twenty years and was finally abandoned during the 1960s. Today there are the remains of buildings around the well, a water tower, three sheds belonging to a wagon factory, and two buildings used as a laboratory and residence.

5.2.2.6 British Ammunition Depot (Figs. 5.98, 5.99 and 5.100)

Narrative

The southern coastal region between the Gaza Strip and Kibbutz Be'eri is a flat region whose terrain was shaped by the nearby Besor stream. The Besor is the lowest point in the landscape, to which numerous channels drain, creating a deeply etched plain. The soil is primarily loess overlaid on eolianite sandstone (*kurkar*) bedrock, a combination that generates unusual landscapes. Among the many names local Arab inhabitants have given the area, one of the most prominent is *nh'abir*—channels formed by rainwater. The area is the meeting point between the Coastal Plain to the north and the desert to the south. This ecotone melds a rare combination of plant and animal habitats and provides encounters between nomadic and farming cultures.

The Landscape

For hundreds of years, this region remained uninhabited by permanent communities and was grazed intensively by Bedouin herds. In the 1930s, a plant for sulfur production

Fig. 5.92 Grove house—Burj Binyamina—polygon

was built: roads were laid out, mines were quarried, buildings were constructed, trees were planted, and railroad tracks were laid, the remains of which are still visible today.

The Bedouin in the area were nomadic until the development of the sulfur plant, living in temporary dwellings based on the availability of water sources and pasture. When the plant was built, they underwent a process of permanent settlement, building brick houses near the plant, digging wells, planting gardens, and developing agriculture and flour mills.

Fearing that the war front was getting closer to Palestine, the British army established defenses against a possible invasion by the German army from Egypt. The British used the sulfur plant to develop a large complex to store aircraft bombs. The existing mining infrastructure, including the narrow-gauge rail line and water facilities, was well suited for use in a military compound. More than 170 warehouses were built in a U-shaped formation to house aircraft bombs for planes stationed at the airport in Gaza, located near the present-day Karni border crossing. The dirt road used to transport sulfur ore to Gaza was paved to facilitate the traffic of heavy trucks carrying bombs and planes. It is still possible to see the V-shaped insignia of the Royal Air Force etched into the surface of the road.

Defining Polygons

The British munition depots are evident in the field, and the cultivation of the surrounding fields has barely affected

Fig. 5.93 Grove house—Khan Zarkaniyya—polygon

them. Plowing tractors sit in the warehouses, and dirt barriers delineating the area jut out of the field. The cultural landscape polygon includes all the remnants of the munitions stores and a network of roads around them. The remains of the munition warehouses are scattered in an area designated for agriculture. The definition of special agricultural areas does not guarantee the preservation of the concrete roads and the ammunition stores. Once the land is prepared for cultivation, the remains of this cultural landscape can quickly disappear.

5.2.2.7 The Workers'Camp in Sodom (Figs. 5.101 and 5.102)

Narrative

"The desert covered her with soft white dust." (From *The Arava House* by Haim Hefer and Amos Kenan).

The Dead Sea Works was founded in 1931 at the initiative of mining engineer Moses Novomeysky, who won the concession for mineral mining at the Dead Sea. They produced compounds of potassium for fertilizer, bromine and

Fig. 5.94 Grove houses

magnesium chloride. The heart of the plant was built along the north part of the sea, but it was not long before the plant was relocated to the southern section of the Dead Sea, which was better suited for creating the shallow pools of water required for evaporation. In 1934, the first workers came from the north part of the Dead Sea in boats and began the plant's initial construction in Sodom. The settlement in the southern part of the Dead Sea was accomplished by the labor unit of the United Kibbutz Movement. For over three years, they paved roads and built dams, ponds and other plant facilities, as well as the camp's living quarters. The existing pier was used by boats, which were the only means of transportation connecting Sodom with the northern Dead Sea. From that point, the workers would continue by car to Jerusalem. Until 1948, the drinking water for the plant came from the Safi oasis in Jordan. The heat, the lack of fresh water, the isolation and the dependence on boats to connect to the rest of the country all contributed to the extremely harsh living conditions in Sodom. Indeed, between 1934 and 1948, Sodom was one of the most isolated communities in the entire country. More than two hundred workers lived in the camp at the base of Mount Sodom, many of whom lived without their family members.

During the War of Independence, hundreds of workers were evacuated, and fighters remaining in the plant in the northern Dead Sea traveled by boat to the plant in Sodom. Between April and November 1948, Sodom was under siege. The siege ended after Operation Lot, when forces broke through to evacuate those remaining. A core group of six workers remained after the war to preserve and maintain

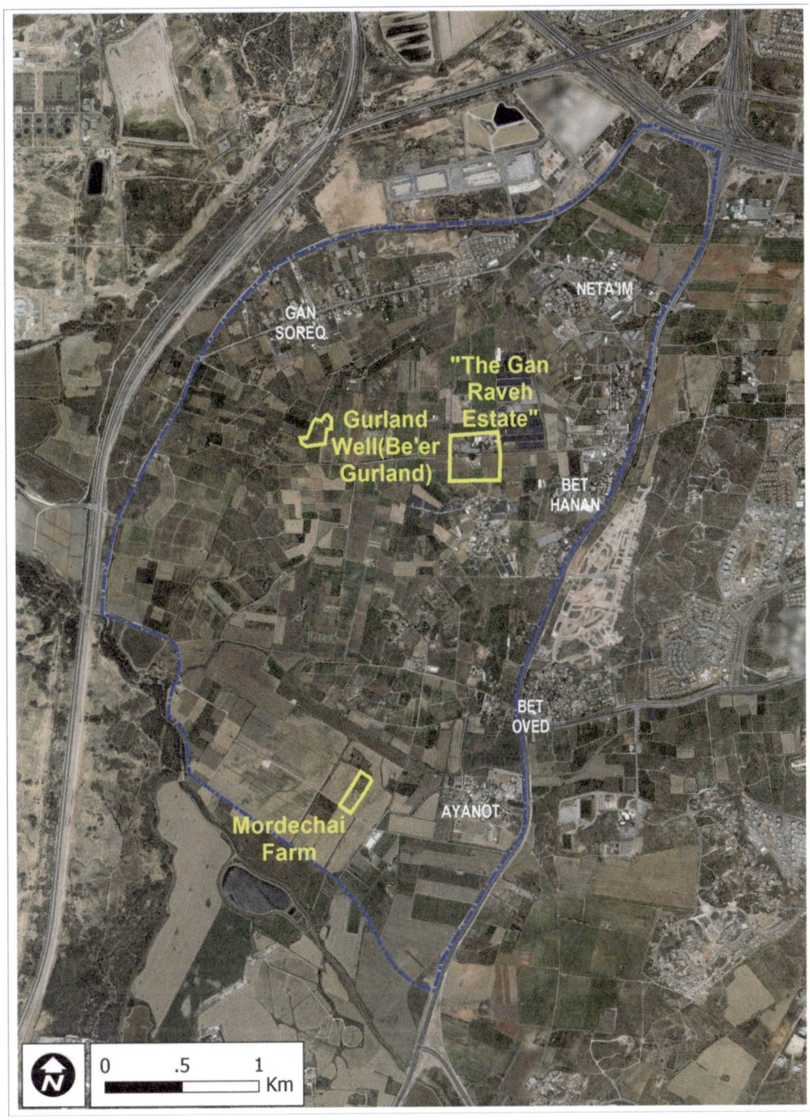

Fig. 5.95 Grove House Estates—polygon

the camp and plant until the formation of the Dead Sea Works in 1952. A road was built between Sodom and Beersheba, and a residence for hundreds of workers was reestablished at the camp. The camp was formally disbanded in 1957. The employees moved to other Negev communities with their families and came to work every day by organized transportation. In 1964, a new plant was built in Sodom south of the original plant.

The Landscape

The remains of the abandoned workers' camp viewed against the backdrop of evaporation ponds and the arid Mount Sodom attest to a chapter in Palestine's industrial history. The wooden shacks where the workers had lived remain in situ. Some workers, mainly those from kibbutzim, lived in tents, the bases of which are still in place. There were several public buildings at the entrance to the chimney caves in Mount Sodom, including a restaurant, a movie theater and a synagogue, which enjoyed the relatively cool air emanating from the caves. The camp also contains the remains of a three-story safe house made of concrete, which housed the clinic, the weapons store and administrative offices.

Defining Polygons

The boundary of the cultural landscape polygon includes the entire camp. The area has not been preserved in any way, and the remains of the buildings are rapidly eroding from

Fig. 5.96 Entrance to *HaAhuza* at the Gan Raveh Estate

years of neglect. There is a real threat to their continued existence. The Dead Sea shore where the camp was built is an artificial pool. It is currently experiencing rising sea level due to the accumulation of salts.

Cultural Anchors

Workers' living quarters, the dining hall, the Colone's cave.

5.2.2.8 Wilhelma (Bnei Atarot) (Figs. 5.103, 5.104 5.105 and 5.106)

Narrative

The Templar movement was a religious movement founded in southern Germany during the second half of the nineteenth century to revive the Protestant Church through deepening spiritual practices. The name 'temple' does not refer to a physical structure; rather, it spiritually expresses the idea of "being and dwelling with God."

Against a background of local difficulties with agriculture in Germany and a rekindled interest in the forgotten Jerusalem, occupied by the Ottoman Turks for over 400 years, they believed that only the return to Jerusalem, in the words of the prophets, would bring change to the troubled world and abolish hunger. In 1854, the "Association for Joining with God in Jerusalem" was established. In 1858, following extensive preparations and a survey expedition, the first Templars arrived in Palestine. The first colony was established in Haifa, followed by a second colony in Jaffa. The nearby Sarona colony was later added as a satellite of the Jaffa colony. With the inauguration of the Jerusalem colony in 1878, the community's spiritual center shifted to the Holy City. A fifth Templar colony was founded in 1902 by the

Fig. 5.97 An abandoned grove house in the Judean Lowlands

second generation of farmers from Jaffa and Sarona who lacked agricultural land due to the urban growth of Jaffa. The colony was established on an area of 890 ha and was called "Hamidiye—Wilhelma." The name "Hamidiye" was in honor of the Ottoman Sultan Abdul Hamid II and "Wilhelma" after Kaiser Wilhelm II, who visited the country in 1898.

The history of the Templar colony echoes the development of modern agricultural practices in Jewish settlements in Palestine. However, it also reflects the relationship between Britain and Germany during the world wars, the upheavals in German society at the time, and the relations between Jews and Germans. During World War II, the British expropriated the Templar properties and transported the community to Australia. The colony became a British base, which was taken over by the Arab Legion after the end of the British mandatory period and was later captured by Israeli forces. After several months of abandonment during the War of Independence, Jewish farmers who were forced to leave their communities resettled the land. They came from Atarot near Jerusalem, as well as Nehalim, Beerot Yitzhak and Ben Shemen. They renamed the settlement Bnei Atarot after the community they had left.

The Landscape

Originally laid out in the form of the Templar Cross, the streets of the colony have been preserved intact. The public buildings are also intact, and although they no longer serve their original purpose, their former uses are still evident. The Templar school is currently used as an administration building for the regional school, and the modern Templar dairy building serves the community administration. The town hall, the most important public building in the Templar colony, is currently used as a sports hall. The town is picture-perfect, as if from a different country in another era. The spacious stone houses and windows with wooden shutters and blossoming window boxes evoke a European village. Fences enclose the yards, and mature eucalyptus trees line the streets, creating a rare landscape in central Israel: a site with an urban fabric that has not changed much

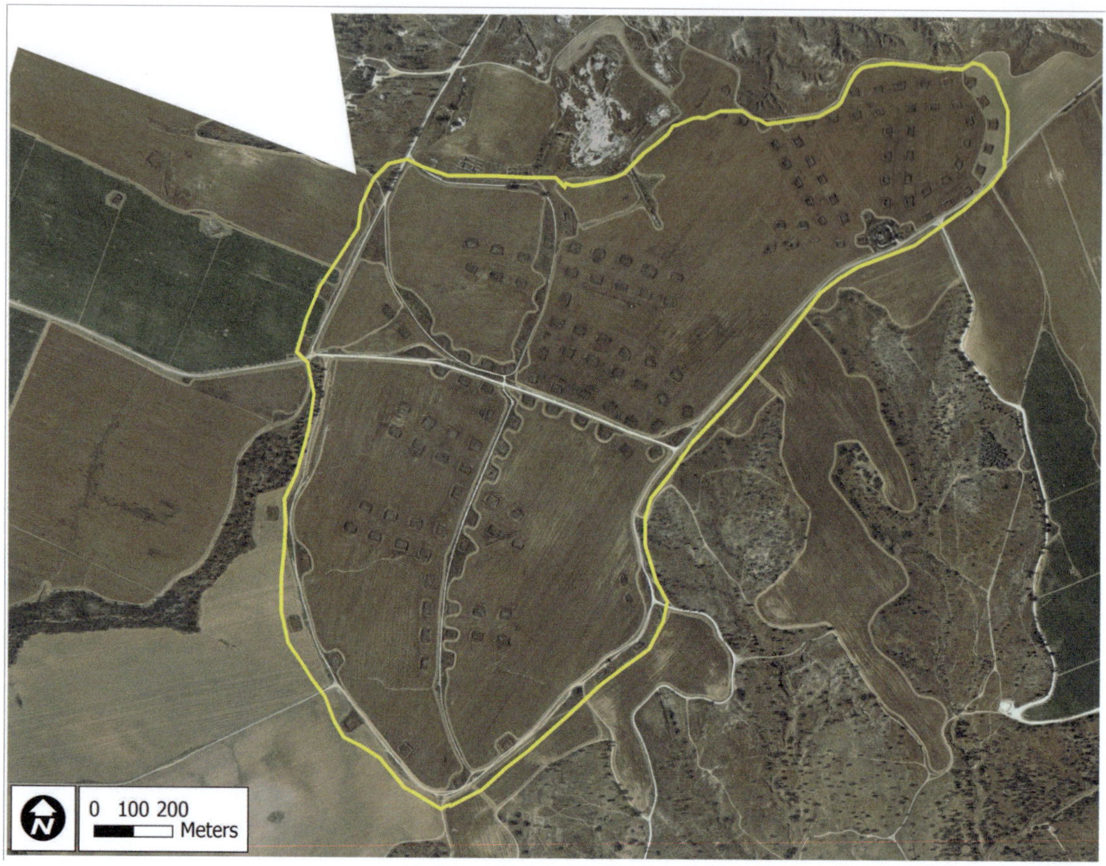

Fig. 5.98 British ammunition depot—polygon

in over a century. The Templar fields are still evident in the landscape, despite the different character of the agriculture. The Templars cultivated grain, planted vineyards and orchards, and raised cows for milk, though today, only orchards dot the landscape. The main road from Wilhelma to Sarona and Jaffa still traverses the fields and is used as a farm road, though a new access road was built due to the construction of the nearby Highway 40.

Defining Polygons

Moshav Bnei Atarot has expanded three times since its establishment, with new neighborhoods built on some of the fields and arable land which was originally part of Wilhelma. Extensions to Highway 40 pass near the community, and the nearby Ben-Gurion International Airport encroached on the margins of the original colony's land. Despite the expansion, it is still possible to identify the original cruciform plan and observe the relationship between the fields and historic streets. This area has been defined as a cultural landscape that should maintain its agricultural component as an integral part of the values. Around the cultural landscape polygon, a minimal buffer was defined, including the farmland and historic roads still in use.

5.2.2.9 *Monasteries in the Judean Lowlands (Shfela) (Figs. 5.106 and 5.107)

Narrative

In the late nineteenth and early twentieth-century, Christian migration from western countries resumed in Palestine, and many orders of Catholic monks built new monasteries across the country outside the traditional holy cities. The Judean Lowlands was an ideal location for these monasteries, partly due to their proximity to the holy places in Jerusalem and Bethlehem, where their orders were based. The monks were attracted to the biblical landscapes of the Judean Lowlands: in the Ayalon Valley, Joshua commanded the sun to stand still (Joshua 10: 12), and the Tzora ridge was where the spirit of the Lord came to the heroic Samson (Judges13: 25).

The Latrun Monastery was founded in 1890. The name is most likely derived from an Arabic distortion of the French Crusader name of the fortress: *La Torun de Chevalier*, the "Small Fortress of the Knights." However, the name could also be derived from the Latin, *latro*, meaning thief. It was believed that this was where the "good thief" was born, one of the two thieves crucified next to Jesus, who repented before his death.

Fig. 5.99 A view of the British ammunition depot in the landscape

The monks belong to the Trappist order, whose members lead an extremely ascetic lifestyle. They maintain a vegetarian diet and try to minimize all talk of worldly matters, thus earning the name "monastery of the silent." The monastery's farm buildings were built around a stone building erected in 1861 and served as an inn for travelers on the new Ottoman road leading to Jerusalem, which passed through Sha'ar Hagai. Olive groves and vineyards surround the monastery, cultivated by the monks.

Just north of the Latrun monastery and Highway 1, on the side of the road to Beit Horon, is the Peace House Inn, belonging to the Beth Haran Catholic order headquartered in Bethlehem. Built in 1927, the building houses the Beatitudes community, a monastery and a hostel for pilgrims. Between 1948 and 1967, it served as the headquarters for UN observers. The remains of a Byzantine and Crusader church were found below the monastery, which sanctified the place as Emmaus. Jesus revealed himself to his disciples after his crucifixion and resurrection in this place. Excavations have revealed remains of a Roman villa with mosaic flooring, a Byzantine church decorated with mosaics from the fifth century, and an adjacent baptistery for immersion. A Crusader church was built on the foundations of the Byzantine church in the twelfth century.

The Deir Rafat Monastery is located on the slopes of the Tzora ridge. The monastery was established in 1927 on the ruins of a Byzantine church. Until 1948, the church also operated an agricultural school. The church walls and ceiling are inscribed with the passage "Hail Mary" in 343 different languages. A statue of the Virgin Mary on the top of the church is visible from a distance.

The Landscape

Most monasteries built in the Holy Land throughout the nineteenth century were located in cities. The Galilean monasteries outside of Nazareth, including Capernaum and

Fig. 5.100 British ammunition depot-detail

Mount Tabor, were built on the remains of Christian holy sites and focused on maintaining the site and serving pilgrims. Rural monasteries composed of agricultural communities were not common in Israel. They are rooted in a European rural model imported to the Israeli countryside. As such, they represent a unique form that is worthy of preservation.

The monasteries of Latrun and Deir Rafat are located on the hillsides that overlook the sources of the Ayalon and Sorek Rivers, which flow over the rolling plains. The stream channels have formed wide valleys over time, with fertile brown soil cultivated with fields, vineyards and orchards, as in biblical times.

The Tzora ridge, covered with pine forests, separates the two monasteries. A scenic road connects the Ayalon River basin with the Sorek River basin and provides expansive views of the Judean Hills, Lowlands and the Coastal Plain.

Defining Polygons

The Catholic monastery Beit Jimal, built in 1881 by the Salesian order, should also have been included in this cultural landscape polygon. However, due to the expansion of the nearby city of Beit Shemesh, the monastery is now surrounded by new city neighborhoods. The monastery and the nearby fields have been preserved, but the value of the cultural landscape has been damaged, as its integrity and authenticity have been compromised. The creation of an open buffer zone that links the three remaining monasteries in the region emphasizes their relationship to a shared "story" in time and space. This may prevent additional harm to these cultural landscapes, as was the case at the Beit Jimal monastery.

5.2.2.10 *A Row of *Tels* in the Judean Lowlands (Shfela) (Figs. 5.108 and 5.109)

Narrative

The Judean Lowlands is a region of soft rolling hills in a transitional area extending between the Coastal Plain in the west and the Judean Hills to the east. During the Miocene period, some twenty million years ago, the sea flooded the western part of the Lowlands and created a landscape with an altitude of 200 meters above sea level. This area is known as the lower plains. The terrain rises to an altitude of 400 meters or more above sea level to the east.

Fig. 5.101 The workers' camp at Sodom—polygon

The Landscape

The region is built on soft chalk, in contrast to the harder limestone and dolomite rock characteristic of the Judean Hills. This rock created hills typical of the Judean Lowlands, enabled the cultivation of the area and served as a suitable raw material used for construction. Its fertile valleys provided easy passage and were the sites of the ancient roads that supported settlements in the region. Important crossroads were located where the high and low plains meet, next to broad arable valleys and close to water. In this area, ancient settlements were built and fortified, then destroyed and rebuilt, until they became prominent *tels* in the landscape.

The higher hills are not suitable for modern agriculture; cultivation is limited to the valleys. The hills are covered with Mediterranean scrub or forests planted in recent decades by the Jewish National Fund. The *tels*, bare and mostly treeless, stand out in this contrasting landscape.

Defining Polygons

Each *tel* has an individual story; these stories are defined by cultural landscape polygons and include the fortified *tel*, cemeteries, and immediate agricultural hinterland. However, the row of *tels* is, first and foremost, the story of the entire complex. This area was surrounded by a buffer zone and included all the *tels* and extensive areas of the Judean

Fig. 5.102 The workers' camp at Sodom

Lowlands. It is possible to define this territory in geographic terms. From every *tel,* it is possible to see at least one other *tel*, and the buffer zone polygon includes these geographic connections. It is also possible to define the buffer zone polygon in cultural terms shared by most *tels* as a complex.

The prophet Joshua described the lands of each of the twelve tribes (Joshua:15). In the description of the border along the lands belonging to the tribe of Judah, forty cities in the Judean Lowlands are listed. The majority of these *tels* have not been identified definitively. However, some have been identified beyond dispute, and many form a long row extending between Tzora in the north and Lachish in the south. Due to their strategic location, the biblical Rehoboam, the first king of Judea, fortified these cities as part of a campaign to protect the kingdom against an Egyptian invasion, as described in the Book of Chronicles, Chapter XI. However, many of the *tels* in the row are not accounted for in the biblical list of cities.

Cultural Anchors

Tel Tzora—the *tel* was not identified by an archaeological excavation but by the preservation of its name: the Arab village built on the *tel* was known as Sar'a. The *tel* has ancient remains, including altars, two tombs, wine presses and cisterns. A tradition in recent years identifies the altar as Manoach's altar, the rock on which he sacrificed peace offerings when the angel of God appeared to him. This tradition also attributes the two tombs to the biblical hero Samson and his father, Manoach.

Tel Beit Shemesh—the *tel* is located in a strategic spot overlooking the eastern opening to the Sorek Valley, one of the roads that link the coast with the Judean Hills. The city of Beit Shemesh is mentioned several times in various biblical texts, though no mention has been found in external sources. The city is mentioned in the Book of Samuel when the Philistines sent the Ark of the Covenant back to Israel, and in King Solomon's list of administrative districts in the

Fig. 5.103 Wilhelma (Bnei Atarot)—polygon

Book of Kings. According to findings from excavations, many buildings were erected in the city during the second half of the tenth century and the beginning of the ninth century BCE, including an underground water reservoir for use during a siege and the oldest metal workshop discovered in the eastern Mediterranean.

Tel Azekah—the *tel* rises one hundred meters above the Elah stream, which flows along its base, making it a natural fortress. The Assyrian King Sennacherib conquered Azekah (701 BCE) and described it as a nest of eagles. When Nebuchadnezzar, the King of Babel, launched his destructive campaign in Judea (586 BC), the cities of Azekah and Lachish were the last to surrender.

Tel Goded—this *tel* has been identified and widely accepted as Moreshet Gat, the city of the prophet Micah. The *tel* was settled during the Early and Late Bronze Age and was resettled as the kingdom of Judea expanded over the Coastal Plain. The settlement was destroyed when Hezekiah's rebellion against Assyria was crushed in 701 BCE and was rebuilt during the Hellenistic period lasting until the end of the Byzantine period.

Tel Maresha—this is the intersection of the roads to Ashkelon, Gaza, Tel Tzafit, Hebron and Jerusalem. Tel Goded and Azekah overlooked the route between Maresha and Jerusalem. The large *tel* was inhabited between the Iron Age and the Roman period.

Fig. 5.104 A Templar building at Wilhelma (Bnei Atarot)

Fig. 5.105 Typical Templar buildings at Wilhelma (Bnei Atarot)

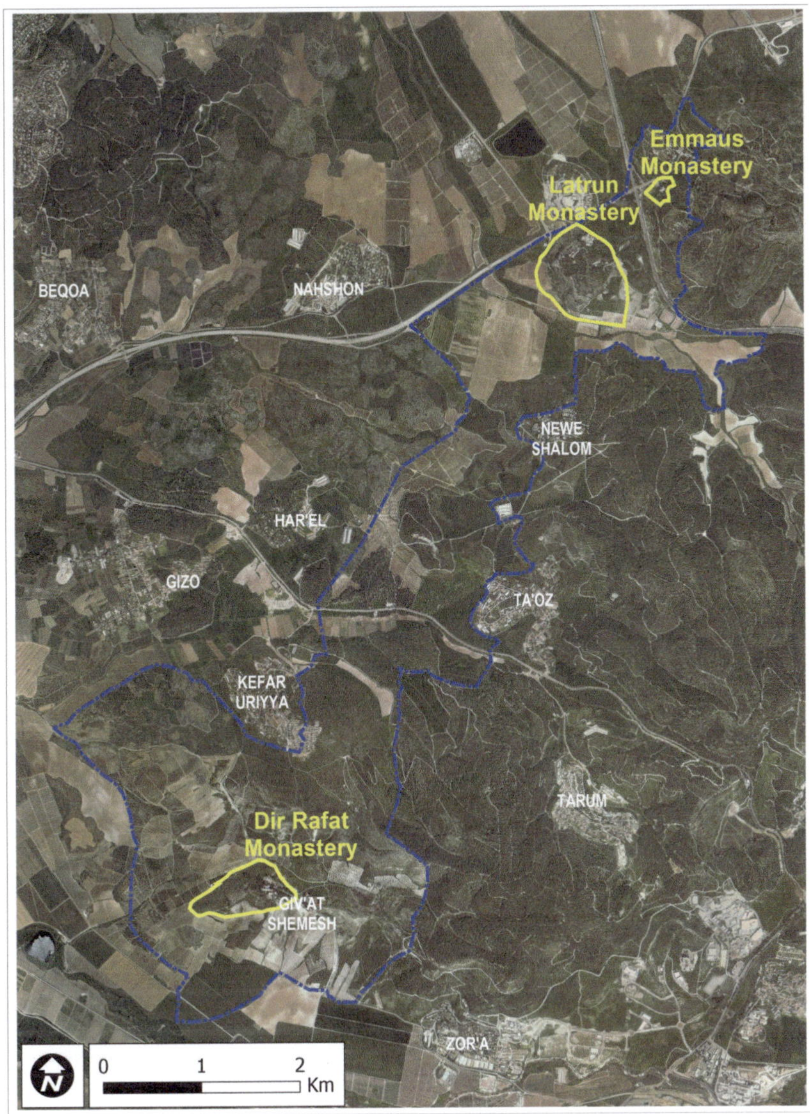

Fig. 5.106 Monasteries in the Judean Lowlands (Shfela)—polygon

Tel Lachish—this *tel* rises over the Lachish stream (Nahal Lachish) stream, along which the central road extends between the Coastal Plain and the Hebron Mountains. The city of Lachish was first mentioned during the Late Bronze Age, including a reference in the Amarna letters. The biblical city of Lachish is mentioned several times in the Book of Joshua, when the five Amorite kings gather to fight in Gibeon. In the Book of Kings, the city of Lachish is mentioned as one of the most important in the Kingdom of Judea. The most significant finding from Lachish at the end of the Judean monarchy (circa 590 BCE) is the "Lachish Letters," discovered during the 1930s, comprising 21 ostraca (inscriptions written in ink on potsherds), written on the eve of the Babylonian conquest. They discuss administrative matters and events of the time. The city was abandoned at the end of the Hellenistic period in the second century BCE.

5.2.2.11 Shikma Park: Rows of Sycamore Fig Trees (Figs. 5110, 5.111 and 5.112)

Narrative

Ficus sycomorus trees (*shikma* in Hebrew) are the most prominent and impressive trees of the Coastal Plain. Sycamore figs are considered local, though, in practice, all sycamore figs in Israel were grown from cuttings. Since the natural pollinators of the sycamore fig are not indigenous to the region, the tree cannot reproduce naturally. The sycamore fig originated in the riverbeds of the tropical savannas

Fig. 5.107 Latrun Monastery

found in East Africa. In its natural habitat, the tree is widespread and reproduces naturally. Sycamore fig trees have been used by man as a source of timber for building, firewood and fruit, and were distributed by humans to the north along the Nile to the eastern and southern shores of the Mediterranean.

Sycamore figs were often mentioned in the Bible, Mishna and the Talmud, as they were used extensively for wood and fruit. The importance of the tree is apparent in the multiple laws concerning the trunk of the sycamore and its history, as it is written: "I was no prophet, neither was I a prophet's son; but I was a herdsman, and a gatherer of sycamore-fruit" (Amos 7: 14), "and cedar as plentiful as sycamore fig trees in the lowlands of the Shephelah." (Chronicles II, 1: 16).

In the past, sycamore fig trees were distributed throughout the country. They are most prominent in the coastal lowlands, especially along the eastern edge of the dunes, where the sycamore figs formed continuous rows. This row lies between the the fertile farmland and the dunes. Farmers also used the sycamore fig to stabilize the dunes and mark a boundary. Throughout history, the roads that crossed the country along its length passed the row of sycamore figs that followed the edge of the dunes.

The historical and environmental connection between the Coastal Plain and the sycamore fig trees is reflected in toponymies, the names of places and sites in the area, including Shikma stream ("Sycamore stream"), Moshav Beit HaShikma ("Sycamore House"), and many more.

The Landscape

The sycamore fig was the hallmark of the lowlands: "An indication of mountainous country is [the presence of] millin; an indication of valleys is the palm tree; an indication of streams is the reed; an indication of lowlands is the sycamore fig tree." (Talmud Tractate Pesachim 53a). Since these words were written, the sand dunes between Tel Aviv and

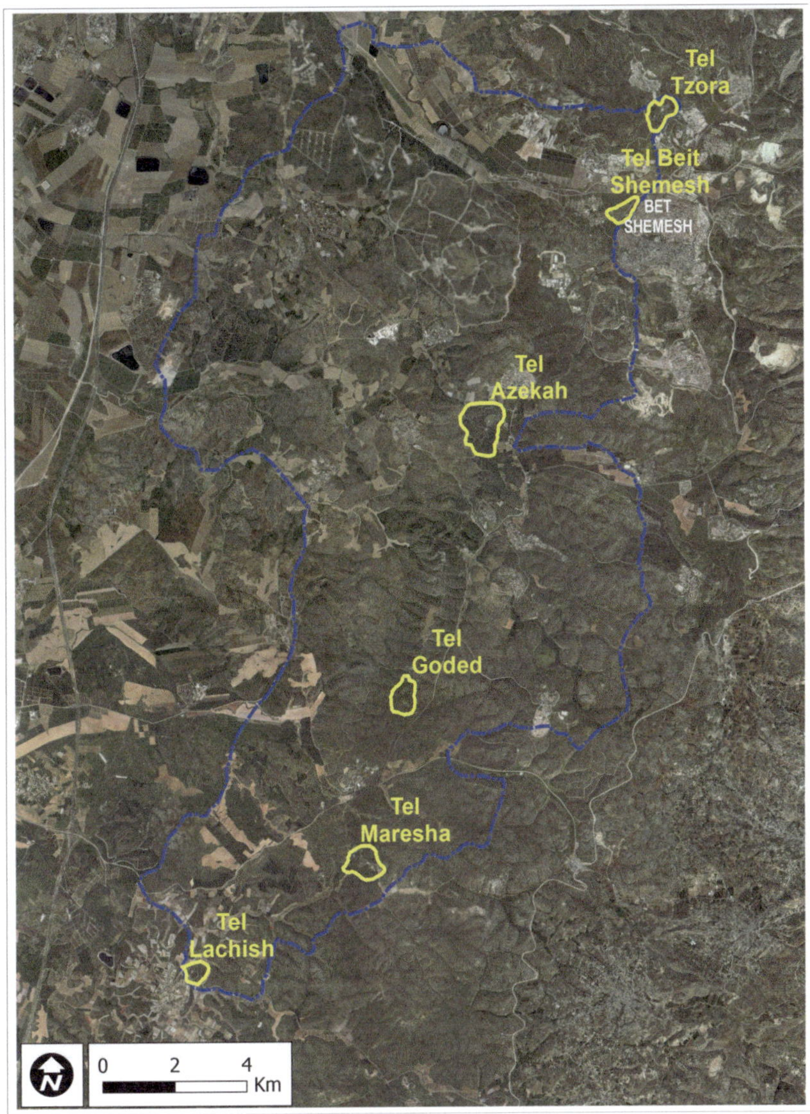

Fig. 5.108 A row of *tels* in the Judean Lowlands (Shfela)—polygon

Ashdod have been built up extensively, and few sycamore fig trees remain. Today, most grow along the coast between Gaza and Ashdod. Most have survived on the eastern edge of the dunes where a relatively significant amount of subsurface water is available between layers of sand and clay. The climate offers a suitable habitat for sycamore fig trees, which have tropical origins. Along this row is an array of orchards in the sands, which have survived in the same habitat despite the lack of maintenance.

Defining Polygons

The row of sycamore figs is a historical pattern of linear plantings that reflects the relationship between landscape and culture, representing local settlement over a significant period and its environmental impact.

The cultural landscape polygons for the rows of sycamore figs are based on the remains along the ancient coastal road between Gaza and Jaffa. Due to the immense cultural and scenic value placed on the integrity of the row, the polygon also includes areas with "gaps" worthy of rehabilitation.

There are numerous threats to sycamore fig trees, and there is a serious concern that they may completely disappear from the open landscape. The primary threats include cutting and removal, decline and lack of reproduction. The region has seen intensive development in recent years, including road construction, expanding towns, sand mining and the growth of agricultural areas in an uncontrolled fashion, causing many of the sycamore figs to be uprooted. In some areas along the row, the sycamore figs are gradually declining. It is assumed that this is caused by the level and

Fig. 5.109 *Tel* Lachish

quality of the groundwater, a reduction in the available water in the soil, and their neglect over the past fifty years. Finally, sycamore fig trees must be planted; they cannot reproduce in Israel without human intervention. The only method of reproduction is from saplings generated from tree cuttings.

5.2.2.12 The Scorpion's Ascent (Figs. 5.113 and 5.114)

Narrative

The name "The Scorpions' Ascent" (Ma'aleh Akrabim) is mentioned in the Bible in the description of the boundary between the tribe of Judah with the Kingdom of Edom: "and he crossed south of the Scorpions' ascent, continued on to Tzin and went over to the south of Kadesh Barnea…" (Joshua 15: 3) The site of the biblical Scorpions' Ascent remains unknown today. There are currently two roads with the name "Scorpions' Ascent:" one is known as the Roman Scorpions' Ascent at Ma'aleh Tzafir, while the other is a modern road built by the British during the mandatory period.

The Tzafir Ascent (the Roman Scorpions' Ascent), called *Ma'aleh Tzafir* in Hebrew and *Nakeb Spaar* in Arabic, was an ancient road three kilometers long that lead down to the Arava along the eastern slope of the Hatzera ridge. The road began in Jerusalem and headed south through Mamsheet, the Tzafir Ascent and Ein Hatzeva. From there, it crossed the Arava and climbed the Mountains of Edom to Basra, an important military point on the eastern border of the Roman Empire.

The pass was likely part of a road paved during the rule of the Emperor Trajan (second century CE) and rebuilt during the early Byzantine period. However, it is possible that the road and the ascent served as trade routes during the Nabataean period. Two strongholds protected the ancient ascent: Matzad Tzafir at the top and Rujm Tzafir at the base. Pottery sherds discovered in these strongholds attest to activity in the third and fourth centuries CE.

Fig. 5.110 Rows of sycamore trees—polygon

During the British mandatory period, the authorities built a steep, winding road along its original historical route. The road connected the police stations in Mamsheet (Kurnub) and Hatzeva. In Operation Lot (November 1948), an IDF force descended the Scorpions' Ascent to aid the besieged Sodom, and four months later, the Golani Brigade used this route to capture Umm Rash Rash (Eilat). Until 1956, when the Ma'ale HaAtzmaut (Independence Ascent) road from Mitzpe Ramon was inaugurated, the Scorpions' Ascent was the only route to the Arava and Eilat.

The Landscape

The Tzafir Ascent and Scorpion's Ascent from the mandatory period follow the eastern slope of the Hatzera ridge until they reach the Tzin stream valley. The highest point of the Tzafir Ascent an altitude of 480 meters above sea level, 300 meters above the Tzin Valley. The pass "slides" along the crest of the tapered sides of the ridges, overcoming the extremely steep slope (40°) using switchbacks and carved steps, preserved for hundreds of meters. Retaining walls were also preserved on the side of the ascent.

The summit of the Scorpions' Ascent is located behind the Hatzera ridge (402 meters above sea level). In the early 1990s, the whole ascent was paved and renovated, though the rusty barrels along its edges have remained as a testament to its past. Overlooks along the road provide spectacular views of the Tzin stream valley and the Roman Scorpions' Ascent.

Fig. 5.111 Rows of sycamore fig trees

Fig. 5.112 A sycamore fig tree

Defining Polygons

During the Roman period, the difference in height between the top of the Hatira ridge and its base was overcome by cutting into the ridge. In modern times the steep topography was negotiated using a winding serpentine road. The cultural landscape polygon includes both engineering methods to navigate the steep slopes. The entire polygon area was declared a nature reserve, and the open landscapes are no longer threatened. However, there are approved plans to mine phosphates south of the area. The buffer zone was outlined to include existing, approved and planned quarry zones immediately adjacent to the Scorpions' Ascent cultural landscape. The plan of the quarry and its rehabilitation plan will have to ensure that they do not damage the cultural landscape.

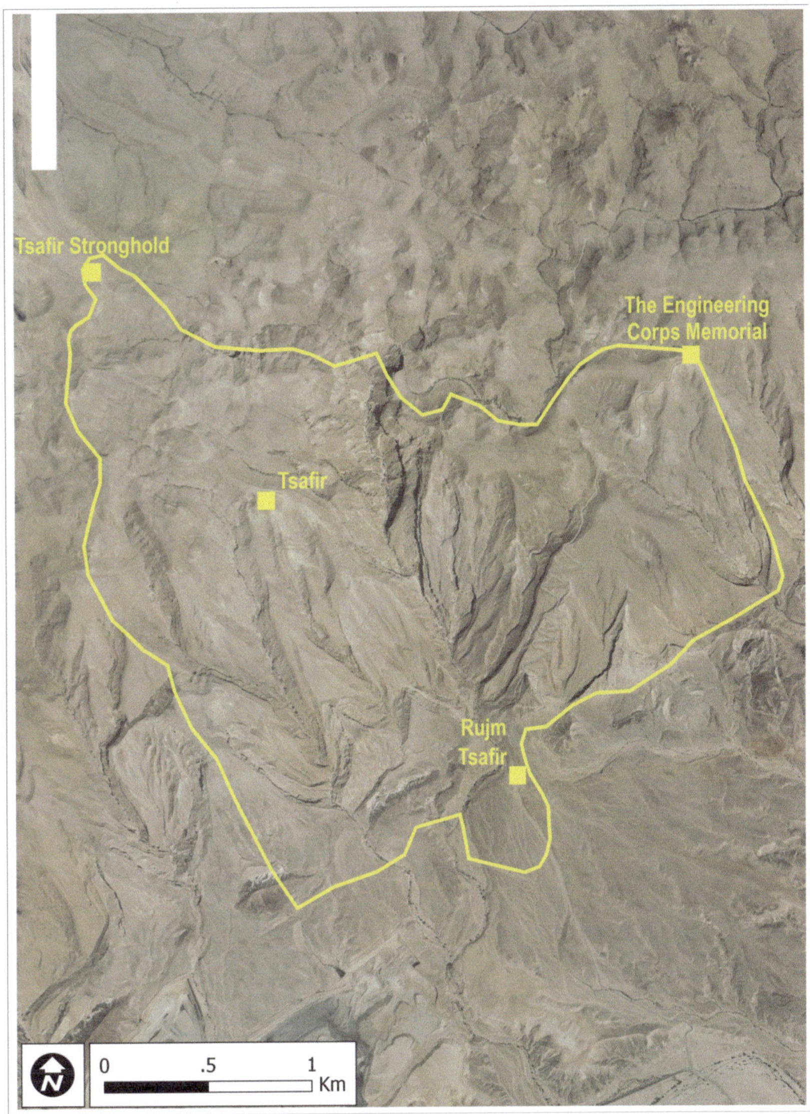

Fig. 5.113 The Scorpion's Ascent—polygon

Cultural Anchors

The Engineering Corps Memorial at the top of the modern ascent, Tzafir Stronghold, Khirbet Tzafir, Rujm Tzafir.

5.2.3 Organically Evolved Landscape: Continuing

5.2.3.1 Olive Groves in the Central Galilee (Figs. 5.115 and 5.116)

Narrative

The olive is the tree that best symbolizes the Mediterranean region. Archaeological evidence indicates that the domestication of olive trees in the Middle East began in the fourth millennium BCE (the Chalcolithic period) when they became part of the agricultural landscape. Olive groves were considered a highly profitable crop, mainly because of the oil produced from the fruit, which was used for illumination and medicine. In describing the virtues of the Land of Israel, the Bible praises the olive: "... a land of olive oil and honey" (Deuteronomy 8: 8).

The cultivation of olives has resulted in many local varieties that are rich in oil. Today, olive groves are irrigated to increase yield. However, in the Arab and Druze villages in the Galilee, where olives have been grown for centuries, the groves are rain-fed, and these make up the majority of olive groves in the country. These Galilee villages not only continue the tradition of olive cultivation but also maintain the lifestyle associated with it. During the harvest season, families go to their olive groves and spend entire days harvesting. The crop is transferred to a local olive press where the oil is produced, generally in return for a portion of the olive oil rather than money.

Fig. 5.114 The Scorpion's Ascent

The Landscape

Due to the rising demand for olive oil, many olive groves have been planted recently. Jewish farmers have joined their Muslim, Christian and Druze neighbors in cultivating olives. Twenty-two thousand hectares of olives have been planted in Israel in every climate zone, from the slopes of Mount Hermon to Ramat HaNegev. Four hundred hectares of olive trees are planted each year.

Defining Polygons

The landscape chosen to represent the continuity of olive culture is in the heart of the Galilee, where olives have been cultivated continuously for centuries, over extensive areas. Throughout the Galilee, it is possible to see ancient olive trees with impressive trunks, some hundreds of years old. Particularly large olive groves grow on the mountain slopes that descend from the Beit HaKerem Valley, as well as within the valley. The name is derived from the area's many olive groves. Other areas with significant groves are located on the southern slopes of Mount Hazon, near the village of Maghar, and between Eilabun and Arabe in the Sakhnin Valley.

The area of olive groves adjacent to the village of Maghar is connected by a narrow strip to the olive groves of En al-Assad village on the outskirts of the Beit HaKerem Valley. The connection point is the "Hanania Farm," a farm established during the 1930s as an experimental farm by the British during the mandatory period. Acclimatization attempts were made on different varieties of fruits gathered around the Mediterranean Basin, and there are some 30 varieties of olive trees in this "Mandate collection." Currently, there is an "Olive Council" operating on-site, with plans to build a museum, tasting rooms, and an archeological garden illustrating the oil production process from biblical times to the present.

Even though the area has preserved the overall fabric of the olive culture in the groves and olive presses, and the way of life associated with the harvest and olive oil production, there are many temptations to abandon olive cultivation for roadside businesses such as gas stations, restaurants and retail stores. These enterprises provide easier and faster income for the price of removing a few trees, which cause little harm to the family's crop yield. The resulting "holes" in the groves mar the continuity and affect the integrity of the

Fig. 5.115 Olive groves in the Galilee—polygon

cultural landscape. To strengthen the cultural landscape of the Galilee olive groves, it is worth investing in treated water. Irrigation can potentially contribute to a fourfold increase in the olive crop, thus reducing the temptation to abandon olive cultivation.

Cultural Anchors

Hanania Farm.

5.2.3.2 Beit Netofa Valley (Figs. 5.117, 5.118 and 5.119)

Narrative

The valley is named after an ancient local settlement (currently known as Khirbet Natif) west of Eilabun. Maimonides explains the origin of the name in his commentary on the Mishna: "Because the land is very wet, it is called Netofa, like water dripping." (*Nataf* means drip in Hebrew.) In Arabic, the valley is called *Sahel Batuff*, the Batuff Plain. In the past, it was also called *Marj al-A'rk,* which means "drowned valley," due to its muddy ground.

As a result of the water that pooled in the valley, the most vital east–west travel artery in the Lower Galilee (*Darab el-Hawarna*) skirted the valley to the west. The road extended from Acre to Tel Hanaton, overlooking the western entrance to the valley. From that point, the ancient road continued to the Yavne'el Valley without traversing the Beit Netofa Valley. To this day, no road crosses the valley.

The National Water Carrier runs along the valley's northern edge in a sixteen-kilometer-long open channel to the Eshkol reservoir at the western edge of the valley. In the reservoir, the water quality is improved and the flow is regulated. From there, the water is piped underground.

Fig. 5.116 Olive groves in the Galilee

The Landscape

The Beit Netofa Valley is the widest valley in the Lower Galilee. The valley is approximately sixteen kilometers long and 4.5 kilometers wide at its widest point, surrounded by mountain ranges. It is enclosed by the Yotvat Range to the north and the Turan Range to the south. Only the western portion has an outlet to the Yiftahel River from which the valley drains to the Tzipori River. There is no drainage in the eastern part of the valley, causing the entire valley to flood during the winter, and in particularly wet years, parts of the valley remain flooded until August.

Current agricultural production in the valley includes vegetables, hay and crops for industry. Wheat and barley are grown at the edge of the valley, where the ground is drier. Only seasonal summer crops are grown in the area flooded in winter, mainly watermelon and other melons. When the crops ripen in summer, the farmers set up summer huts on the edge of the fields, with many spending days and nights watching the fields. It is one of the traditional ways of life still practiced in the valley. The residents of nearby Arab villages cultivate the valley in small plots, creating a spectacular mosaic viewed from the surrounding hills. The colors change with the seasons and vary from year to year, depending on the amount of rainfall and the length of the flood period.

Stone walls separate the plots and other uncultivated areas. Many rare plants have survived due to flooding and cultivation conditions, and several subspecies are endemic to the Beit Netofa Valley.

Defining Polygons

The cultural landscape polygon of the Beit Netofa valley was defined within the area of traditional Galilean agriculture, characterized by small plots in which a large variety of crops are grown. Traditional stone walls enclose each plot. To maintain the integrity of the cultural values expressed in the form of the valley landscape, it is advisable to preserve the valley's drainage, maintain the cultivation of the land and the plot parcellation, prevent paving of additional roads perpendicular to the fields and refrain from building (including the construction of farm buildings). By its very nature, the valley is a low area seen from the surrounding

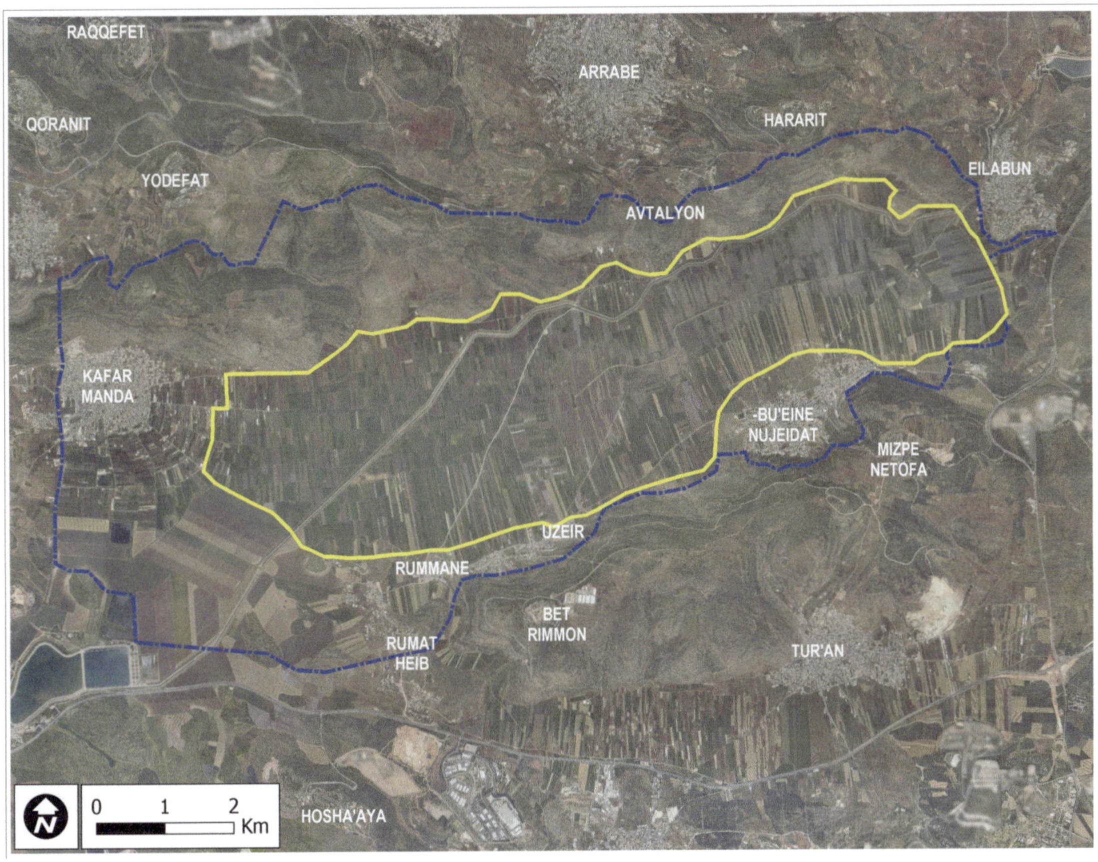

Fig. 5.117 Beit Netofa Valley—polygon

vantage points. The buffer polygon is defined by the ridge line surrounding the valley. Construction and development plans for the area, as well as detailed plans for afforestation, should take into account the changes they may cause to the valley's "sense of place." The proposed plans' effects on the valley must be presented to the local planning authority as part of the approval process.

5.2.3.3 Irrigated Agriculture in the Harod Valley (Emek Harod) (Figs. 5.120, 5.121 and 5.122)

Narrative

The Harod Valley and the Beit She'an Valley are located in a relatively arid area, albeit with rich, lush springs. The springs flow from the the western portion of the valley, at the foot of Mount Gilboa, from the aquifer below the mountains of Samaria near the cities of Jenin and Nablus. The water flow in the valley springs generally amounts to 130 million cubic meters of water annually. Until the twentieth century, the spring water flowed over the surface undisturbed. Most of the water drains into the Harod River, the primary river in the region, and it floods large parts of the valley during rainy winters.

The land purchases made by the Zionist movement beginning in the early twentieth century radically transformed the valley landscape. The Jewish settlement in the Harod Valley exploited the natural water resources until they were exhausted in the 1960s. The springs were diverted for irrigation, and groundwater was pumped, reducing the flow of the natural springs. The availability of water in the Beit She'an Valley region brought about the rapid development of irrigated agriculture in the region. Intensive crops require irrigation due to the very high evaporation rate, which is one-third higher in this region than in the Sharon Region. The springs are found at higher elevations, allowing most of

Fig. 5.118 Beit Netofa Valley

the valley's fields to be irrigated by open gravity-fed aqueducts, one of the unique agriculture characteristics in the region.

The Landscape

The new settlement patterns changed the landscape of rain-fed agricultural fields and marshes before the Zionist period to fields that produced more than two crop yields a year. Aquaculture was also added. The kibbutzim in the valley introduced large-scale agriculture extending over an area of some 12,000 ha. Approximately three-quarters of the agricultural land produces field crops and vegetables. The rest of the area consists of groves (mainly date groves), fishponds and reservoirs.

Several of the largest springs continue to flow in the valley, creating attractive brooks, until the point where the waters are diverted for irrigation. The largest and most important springs are Moda, Shokek, HaKibbutzim stream, Harod and Amal.

Defining Polygons

The cultural landscape polygon includes all of the Harod Valley. The Harod stream is in the center, with fishponds and cultivated streambanks running along its length. The southern bank extends to the foot of the Gilboa Mountains, and the northern bank extends to Yissachar Heights (Ramat Yissachar). When viewed from the top of Mount Gilboa, this landscape reads as a single, coherent unit.

5.2.3.4 HaNadiv Valley (Figs. 5.123 and 5.124)

Narrative

HaNadiv Valley encapsulates the essence of the history of the moshavot from the First Aliyah period and the assistance they received from Baron Edmond de Rothschild. Between 1882 and 1904, thousands of Jews came to Palestine, mainly from Eastern Europe. Many wished to escape the violent pogroms sweeping the region, though there were also many idealistic groups who dreamed of

Fig. 5.119 Beit Netofa Valley in the summer

establishing agricultural settlements. It was within this context that a number of settlements in HaNadiv Valley were founded, including Zichron Ya'akov, Binyamina, Pardes Hanna and Givat Ada.

The new settlers had little agricultural knowledge and nearly failed to realize their vision. Baron de Rothschild established a framework to support and supervise the new moshavot. Although this framework caused serious friction between the settlers and the Baron, there is no doubt that his contribution was critical in developing and stabilizing these communities. In his later years, Baron de Rothschild transferred responsibility for the moshavot to the Palestine Jewish Colonization Association, although he remained very connected to the place and its communities. After the death of the Baron and his wife, their remains were brought to Israel, and they are now buried in the Baron's garden in Ramat HaNadiv (see Sect. 5.2.1.4).

The Landscape

HaNadiv Valley is an agricultural area with outstanding scenic qualities: small plots that are narrow and long laid out perpendicular to the roads and streams. The flowering trees in winter and the warm hues of the grape leaves in autumn create a unique and varied appearance throughout the year. The area's complex hydrological system centers on the Taninim stream and includes natural springs and hewn cisterns.

Defining Polygons

The HaNadiv Valley cultural landscape preserves traditional agriculture. Without protection, industrial agriculture would overtake the area and destroy the local characteristics. The polygon includes agricultural areas divided into small plots common in the older moshavot and plots established by more recent communities.

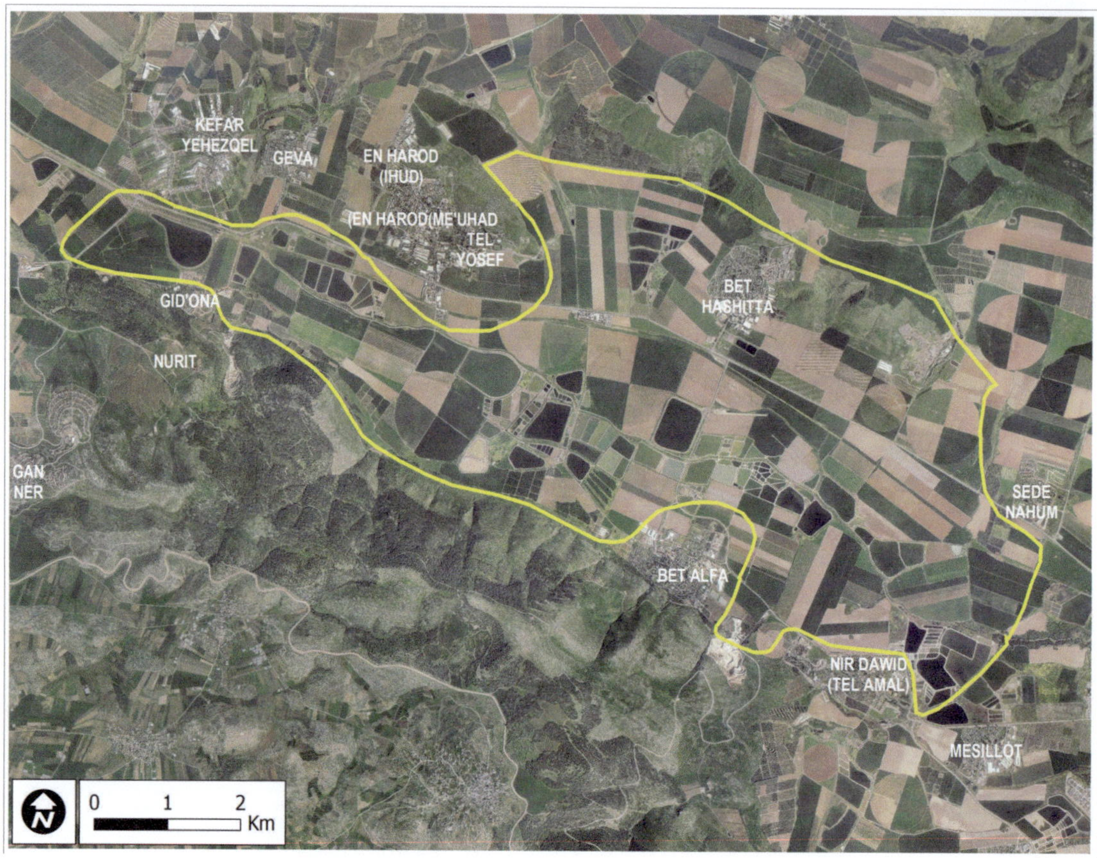

Fig. 5.120 Irrigated agriculture in the Harod Valley—polygon

5.2.3.5 The Groves of the Sharon Plain (Figs. 5.125, 5.126 and 5.127)

Narrative

Large orange groves already existed at the end of the eighteenth century, particularly around Jaffa. By the mid-nineteenth century, citrus was exported from the Jaffa port. These were known as the famous "Jaffa oranges," which are the *shamouti* variety, a mutation of the local oranges which proved to be juicy, exceptionally tasty and with an extensive shelf life.

Jewish farmers initially started cultivating citrus to produce citron or *etrog,* one of the four species used in Jewish ritual during the holiday of Sukkot. The first citron orchard was planted around Jaffa in 1842. By the end of the nineteenth century, Jewish farmers began growing oranges in the Sharon region and the southern Coastal Plain. A turning point occurred in the early twentieth century when wealthier Jews immigrated to Palestine and invested in planting orange groves. The groves farmed by Jews thrived after World War I. The red loam soil of the Sharon region provided ideal conditions for citrus production, and the area was transformed into a forest of groves. These groves replaced the Tabor oak forests of the Sharon, the remains of which can still be seen near Hadera, close to the ruins of the Circassian village, in Pardes Hanna and around Elyachin.

The citrus industry was the first agricultural industry to employ modern methods. Saplings were prepared in nurseries and the soil was treated with manure and fertilizer. Packing was also done professionally: rail lines accessed the packing houses, avoiding the need for transport by camels. The groves employed thousands of immigrants. Until 1936, one out of every four people in the Jewish community made their living from the citrus industry.

Fig. 5.121 Overlooking the fishponds of the Harod Valley

Each grove had a well (see Sect. 5.2.2.4), a grove house and a packing house (see Sect. 5.2.2.5), which were an integral part of the citrus grove workers' lifestyle. The Coastal Plain's groves, particularly the Sharon groves, have been the subject of stories, books, poems and paintings. Writers and artists, including S. Yizhar, Benjamin Tammuz and Nahum Gutman, have described the landscapes in words and paint.

Changes in the economy impacted the local citrus industry. The land of the Sharon became prime real estate property and water became a precious commodity. In addition, other countries in the Mediterranean basin joined the competition, and from an economic perspective, it was no longer profitable to grow citrus in the Sharon. The groves migrated south to the northern Negev, where recycled water was used for irrigation, and the culture of the Sharon citrus groves was nearly forgotten.

The Landscape

The citrus groves extend from Hadera south towards the Itzhak stream. They are visible from the top of Tel Zror, which overlooks the area. The remains of the Ottoman railroad appear as a berm that follows the channel of the Itzhak stream. The rail line was constructed during World War I to reach the Sinai Peninsula.

The Zeita Marsh Reserve is located at the northeastern edge of the area. This is one of the few remaining Hadera marshes in the area; other marshes were purchased from the Arab owners to provide land for citrus production during

Fig. 5.122 Irrigated agriculture in the Harod Valley

the beginning of the Zionist settlement. A winter pool remains on the site, preserving the rare Israeli winter pool habitat.

Defining Polygons

The Zeita Valley, near Hadera, is a "nature reserve" of groves. These groves extend over some twenty-two square kilometers and are the last major concentration of active citrus groves in the Sharon and Coastal Plain area. There is national cultural significance in preserving the region's agricultural landscape of groves. Various tools are available to ensure desirable agricultural crops in a particular region; however, Israel has yet to integrate tools for this purpose, and local tools have not been utilized.

5.2.3.6 Bat Shlomo (Figs. 5.128, 5.129 and 5.130)

Narrative

The development of Zichron Ya'akov in the late nineteenth century encouraged Baron de Rothschild to seek additional land for settlers. The land of Um Al Jamal, an hour and a half donkey ride northwest of Zichron Ya'akov, near the Tut stream, was acquired in 1886, and shortly thereafter, young men from Zichron Ya'akov began cultivating the land. Two years later, a new moshava called Bat Shlomo was established, named after Baron Rothschild's uncle.

To lower costs, residents of Bat Shlomo received most of their services in Zichron Ya'akov, and in many ways, it

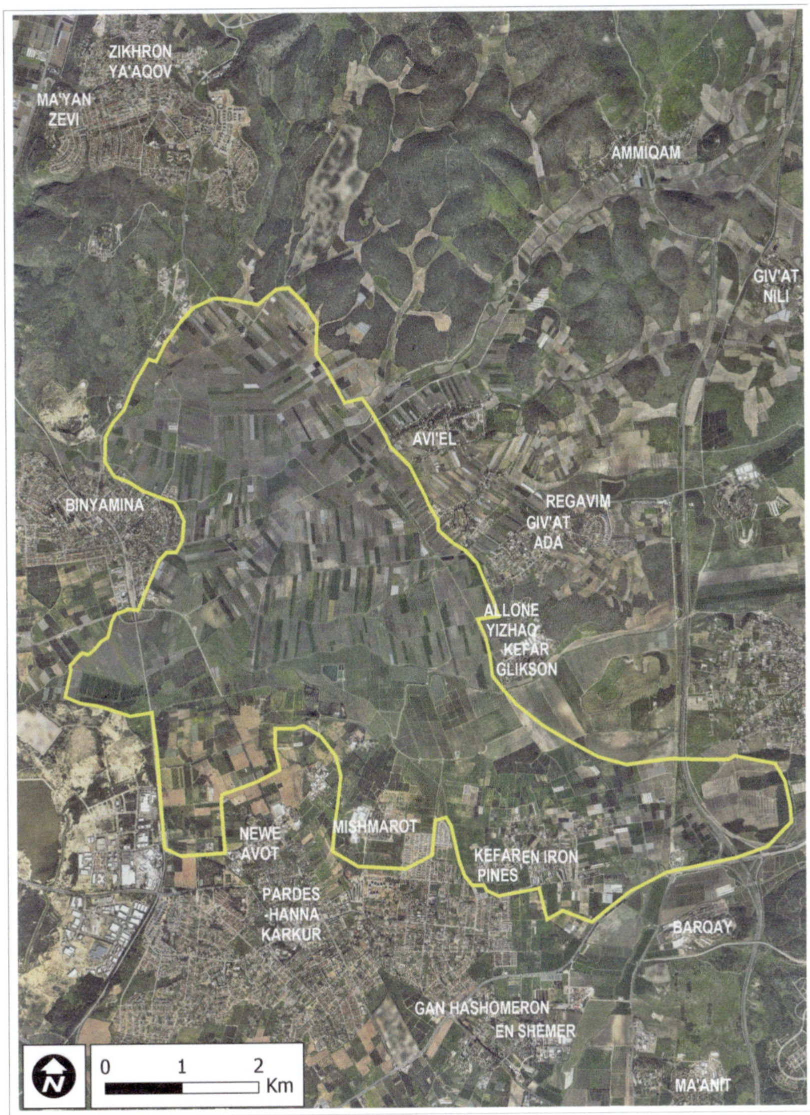

Fig. 5.123 HaNadiv Valley—polygon

remained a satellite community of Zichron Ya'akov. Because of the religious character of the population, residents built a synagogue and a ritual bath. A few years later, pressure from residents resulted in the construction of an independent school.

With the establishment of the moshava, the Baron's representatives tried a new system to increase the independence of the new settlers. The newcomers were carefully selected, and residents were given a relatively large parcel of land (22 ha), long-term loans, tools and livestock. Despite these incentives, the development of BatShlomo was slower and more modest than Zichron Ya'akov and other First Aliyah communities. A shortage of fertile land in the area caused the number of farmers to remain static and eventually decrease as residents left to join other communities.

The Landscape

Bat Shlomo was built south of the Wadi Meelek Road. The community was laid out along one street, flanked by houses on both sides. The houses were built in a similar style, with yards separated by walls. The outer walls were constructed of stone and joined together to form a wall surrounding the community. Every yard had a small residential building, several farm buildings and an outdoor toilet.

In the 1950s, a new immigrant community was built north of the Wadi Meelek Road. Unlike most First Aliyah

Fig. 5.124 A vineyard in HaNadiv Valley

moshavot, where the new construction merged with the older structures and broke the original pattern of the plots, the original plan of Bat Shlomo was preserved. The new homes were set along the historic street and the original pattern of the fields was maintained behind them. Most of the historic buildings have undergone changes and adjustments to update them to modern living standards. However, it is still possible to identify the original core, which includes 14 residential houses and public buildings (the ritual bath, a synagogue, a security building and a public house), and the adjacent farm plots. The proximity between the single main street and the built heritage assets along the Tut stream and the natural forest of Mount Horshan to the south can be seen from every street corner, creating a fascinating encounter between the natural and built environments.

Defining Polygons

The expansion of the Wadi Meelek Road (Route 70) passes north of the moshava, and the construction of a deep ditch cut the community off from its surrounding environment. This construction damaged the historic road yet formed a clear boundary between the original community and the later addition that bears the same name. The main street is the heart of historic Bat Shlomo and is an outstanding and authentic example of the atmosphere and landscapes created during the First Aliyah period. The cultural landscape polygon includes the hill on which the moshava was established, the built-up area and the adjacent farmland.

The buffer defined around the polygon encompasses the plots adjacent to the houses to the south, east and west. Furthermore, it includes all the agricultural land between the

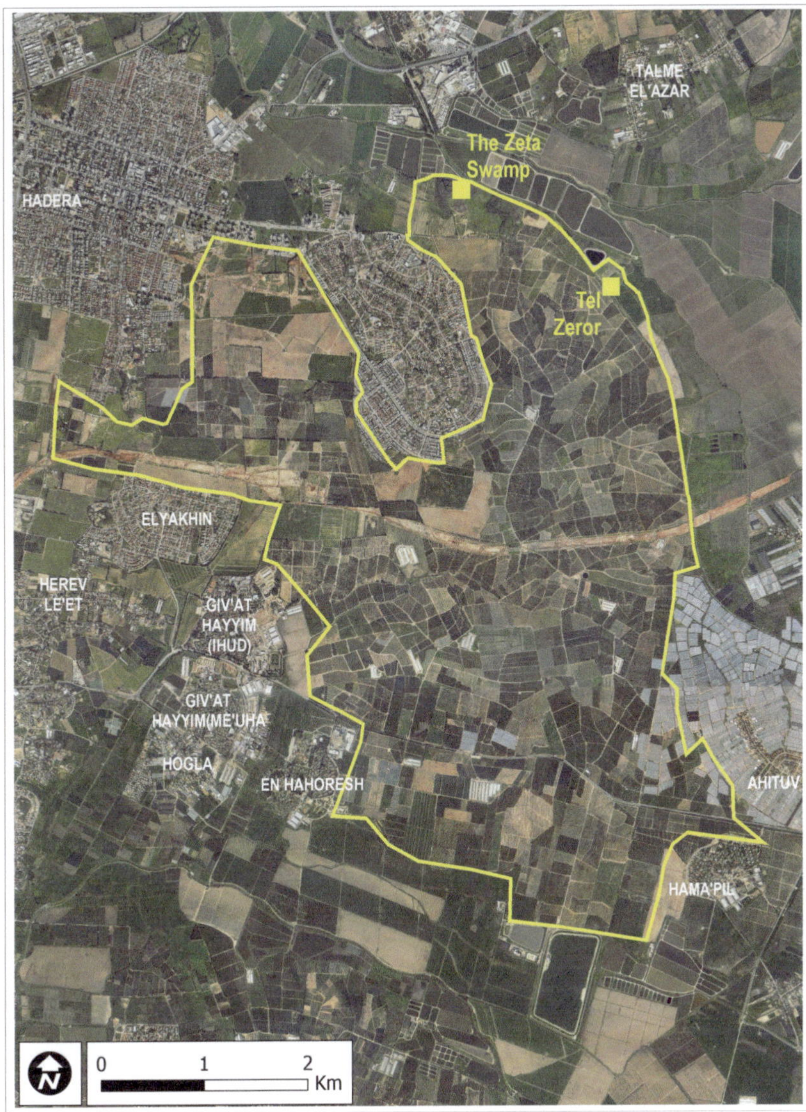

Fig. 5.125 The groves of the Sharon Plain—polygon

stream, the historic community, and the nearby natural areas south of the Tut stream.

The buffer zone has already seen the construction of several large agricultural buildings and infrastructure facilities. Uncontrolled construction of other buildings would cause considerable harm to the integrity of the cultural landscape and undermine the relationship between Bat Shlomo and the natural surroundings that lie south of the community, similar to the damage caused by the road expansion to the north.

5.2.3.7 Mikve Israel (Figs. 5.131, 5.132 and 5.133)

Narrative

Karl Jacob Netter, a representative of the Alliance Israélite Universelle in Paris, believed that "only through working the land would Israel be redeemed." He proposed the establishment of an institution "where Jewish children would learn to work the land, thus enabling the desert land to bloom once again." On April 5, 1870, the Ottoman Sultan

Fig. 5.126 The groves of the Sharon Plain

granted a concession that allowed this idea to be realized, paving the way for establishing the first Jewish agricultural school in Israel. Its name, Mikve Israel, meaning "Hope of Israel," was based on a verse from Jeremiah 14: 8, "Hope of Israel, savior in time of trouble…" Indeed, the establishment of this institution played an important role in the rebirth of modern Israel. This was the first school to train young men and women in agriculture, and its graduates were among the founders of other settlements, including Gedera, Mazkeret Batya and Kfar Netter. In 1882, Netter brought 28 orphans who had been victims of the Russian pogroms to the school. At that time, the school served as a base for members of Bilu, a movement for the agricultural settlement of Israel. Until the establishment of the Volcani Center and the Faculty of Agriculture in Rehovot, Mikve Israel was also the main agricultural research center in Israel and a major center for the acclimatization of imported plant varieties.

The Netter house was a meeting place for the cultural elite of the Jewish community, including Bialik and Ahad Ha'am.[4] The entrance gate to the school was the setting of a brief meeting between Theodor Herzl and Kaiser Wilhelm in October 1898, an event commemorated in the famous photo taken by David Wolfson. The first agronomy books written in Hebrew were also produced in Mikve Israel, and it was the site of the country's first modern winery. The winery's cellars served as a training place for Haganah forces, and the school served as their base during the riots of 1920 and the Arab revolt of 1936–1939. During the 1948 War of Independence, the school was used as a military outpost on the road to Jerusalem. Several weapons were developed by one of the teachers, David Lebovitz, including the "Davidka" mortar.

Today, Mikve Israel is a state school that enrolls more than 1000 students, including junior high, high school and a boarding school. The school has several majors besides agriculture, including a nature studies division, theater and environmental studies.

[4] Chaim Nachman Bialik and Ahad Ha'am were leading literary figures in Palestine at that time.

Fig. 5.127 Orange groves

The Landscape

After years of physical neglect, a plan to renovate the Netter house was proposed in the early 1970s as part of a master plan to rehabilitate the area. Since then, several historic buildings have been partially restored, including the administrative building, the Netter house, warehouses, the winery, and the synagogue. The site also serves as the headquarters for the Council for the Conservation of Buildings and Sites in Israel.

Construction for a botanical garden began in 1929, which includes rare and impressive specimens such as the *Ficus benghalensis* (Bengali ficus). The avenues of *Washingtonia* palms that mark the historic entrance to Miqve Israel from the road linking Jaffa and Jerusalem, are scenic assets that can still be seen today. Verdant fields surround this historic fabric and orchards cultivated by the school's students.

Defining Polygons

The farmlands belonging to the Mikve Israel agricultural school are protected by a law that sets out the specific terms and conditions required to rezone the land. Any changes to the zoning that are not for educational purposes require the consent of three ministers. Despite this special protection, there is no explicit protection against rezoning the open spaces within the school property or the surrounding agricultural landscape to support the school's mission despite its potential impact on the cultural landscape. The purpose of delineating the cultural landscape polygon to include all the school lands and areas adjacent to the Jaffa-Jerusalem Road (Route 44) is to prevent damaging the integrity of this landscape.

The rest of the adjacent farmland south of Highway 44 will be included in Ariel Sharon Park, which has been

Fig. 5.128 Bat Shlomo—polygon

marked as a buffer to the cultural landscape polygon. These areas should remain open to avoid making the agricultural school an isolated green island surrounded by a built-up urban environment.

5.2.3.8 Deciduous Groves in the Ya'afuri Valley (Figs. 5.134, 5.135 and 5.136)

Narrative

In the mid-nineteenth century, many farmers in the northern Golan Heights raised silkworms. The industry had declined by the beginning of the twentieth century. Only a few mulberry trees that fed the larvae remained from that period. The shift towards agriculture in the area occurred in 1945, when Sheikh Salah Tarabi of Majdal Shams planted the first apple orchard in Al-Marj. The high areas of the northern Golan are particularly suitable for growing deciduous fruit trees that require a cold winter, especially apples, as the winter climate in the Golan provides the necessary cool temperatures. The relatively mild summers ensure fruit of high quality and long shelf life. The high altitude (over 900 meters above sea level) increases the amount of violet and ultraviolet radiation, providing a more colorful fruit.

The Druze villages in the Golan did not gain ownership over large tracts of land. Most northern Golan Druze families farm small plots are between 0.3 and 0.2 ha. As a result, the

Fig. 5.129 A peach orchard at Bat Shlomo

Fig. 5.130 The view of Bat Shlomo from Mount Horshan

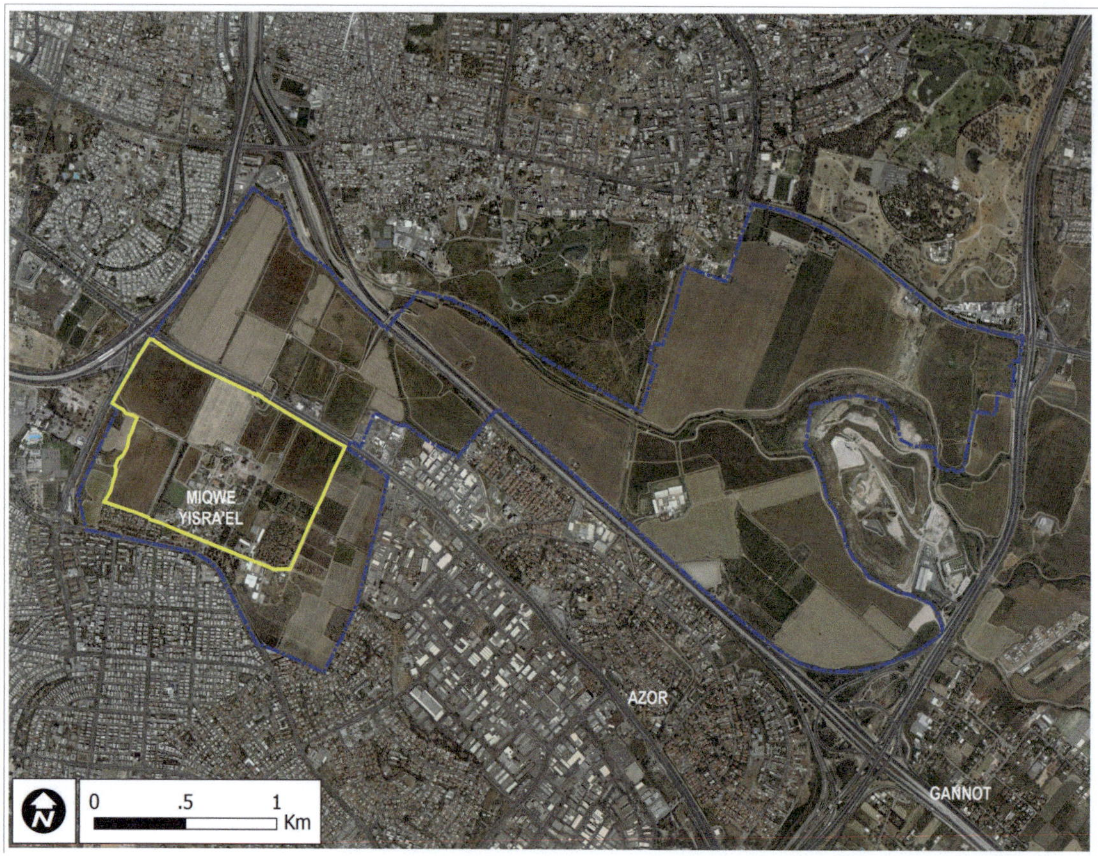

Fig. 5.131 Mikve Israel—polygon

deciduous fruit trees industry has had the greatest economic success. The Druze also grow cherries, grapes and pears, but it seems that the apple known in Israel as the "Hermon apple," is the most profitable crop. Apple cultivation, production and marketing have become modern and sophisticated, and the industry has made an important contribution towards raising the standard of living for the population.

The Landscape

The Ya'afuri Valley spans both sides of the Sa'ar River, at the foot of Mount Hermon. Birkat Ram (Ram Lake) is located in the southern part of the valley. It has an elliptical shape measuring 850 meters long and 600 meters wide. The lake is a "freshwater lens," that is, exposed groundwater, and is used as a reservoir for irrigation. A kilometer north of Birkat Ram is the tomb of Nabi Ya'afuri, a place sacred to the Druze in the Golan Heights, and the name of the surrounding valley was taken from this sacred site. The valley is filled with grapevines, apple orchards, and other deciduous fruit trees. Druze farmers are attempting to expand their orchards beyond the slopes of the fertile Ya'afuri Valley. To solve the problem of irrigation on the slopes and in the Ya'afuri Valley, farmers have built large round metal pools painted in primary reddish-brown colors to collect rainwater. Slopes covered with orchards interspersed with pools are the hallmarks of the region. During the winter, snow covers the area for a few days each year, creating a spectacular view of Mount Hermon.

Defining Polygons

The cultural landscape encompasses the cultivated area between the villages of Masade, Ein Qiniya and Majdal Shams. It includes the anchors of Birkat Ram and the tomb of Nabi Ya'afuri. The valley and the surrounding mountain slopes can be observed from accessible vantage points nearby, from Nimrod, from the road ascending from Masade to Majdal Shams and the slopes of Mount Hermon. To preserve this continuous agricultural landscape, it is necessary to ensure that expanded cultivation of the orchards does not detract from the essence of the landscape values. For example, it is worth considering restrictions on the height of terraces, their design and the size of agricultural structures,

Fig. 5.132 Mikve Israel

especially for roof design and covered parking areas linked to them. Concentrations of agricultural buildings can be placed adjacent to the Druze communities, thus limiting their environmental and visual impact.

Cultural Anchors

Birkat Ram (Lake Ram), Nabi Ya'afuri (tomb of the Ya'a-furi prophet).

5.2.4 Associative Landscapes

5.2.4.1 Tombs of the Righteous (*Tzadikim*) in the Galilee[5] (Figs. 5.137, 5,138 and 5.139)

Narrative

The Upper Galilee has one of the highest concentrations of tombs attributed to the Righteous (Tzadikim) in the country. In a relatively small area of several hundred square kilometers, a few dozen tombs can be found, which, according to Jewish tradition, are associated with rabbinic figures from the period of the Mishna and Talmud (*Tannaim* and *Amoraim*).

The history of these tombs and their early identification is unclear. During the Late Roman period, in the aftermath of the Bar Kokhba revolt, most of the remaining Jewish population moved to the Galilee region. Many figures mentioned in the rabbinic literature were active here. However, the historical documentation of these graves occurred much later. Rabbi Benjamin of Tudela, who visited Palestine in the second half of the twelfth century, is one of the first to mention the presence of these sacred tombs in the eastern Galilee. Later, many more descriptions of the tombs emerged during the Late Islamic period. They became prominent pilgrimage sites for both Jewish and Muslim residents of the Galilee and Jewish pilgrims who came from Middle Eastern countries. Visits to these sites were part of a *ziara* (an Arabic word for visit, meeting or pilgrimage) tour to holy places, which became increasingly common during this period. Today, pilgrimage to the tombs of the righteous is a growing cultural phenomenon, with millions of visits per year. It is estimated that 30% of Israel's population visits

[5] This section was written by Nurit Lissovsky.

Fig. 5.133 The Bengali ficus in the historic courtyard of Mikve Israel

these tombs at least once a year. 15% of the visitors are secular visitors who come for spiritual reasons or as part of a larger phenomenon of cultural tourism and general curiosity.

The Landscape

Until the establishment of Israel, many tombs attributed to the Righteous (Tzadikim) were only marked with piles of stones, caves, pillars and ancient burial caskets. Only after the establishment of the State of Israel did the Ministry of Religious Affairs begin designing these tombs. During this period, appropriate signage was installed, and the tombs were built in a unique style that was adapted to the dramatic landscape of the Galilee. The landscape surrounding the tombs also changed during that period. Mountains that had been bare due to deforestation and centuries of overgrazing were replanted, mostly with conifers.

During the 1970s, the development and discovery of holy sites in Galilee continued, primarily initiated by individuals and associations concerned with the discovery, identification and development of sacred sites. These included the Commission for the Restoration of Holy Places in Israel (*Mishmeret HaKodesh,* "The Holy Watch") and the Committee to Save Ancient Tombs in the Land of Israel, established in the 1980s. These groups set up concrete "tents" over the tombs and completely changed their appearance.

The growing popularity of Jewish pilgrimage to the tombs of the righteous is reflected in the proliferation of new sites, the construction of roads, the preservation, restoration or rebuilding of structures, the commercialization, as well as the addition of services and activities nearby. Some of the sites have undergone such massive development that it is difficult to identify the "spirit of the place" that gives the site its importance in the eyes of the visitors.

Defining Polygons

One of the most important characteristics of the tombs of the righteous in the Galilee is their location in the open landscape rather than in the cemetery of a historic city. The change in atmosphere that occurred as part of the journey from one's home in a town or city to a marked place set in

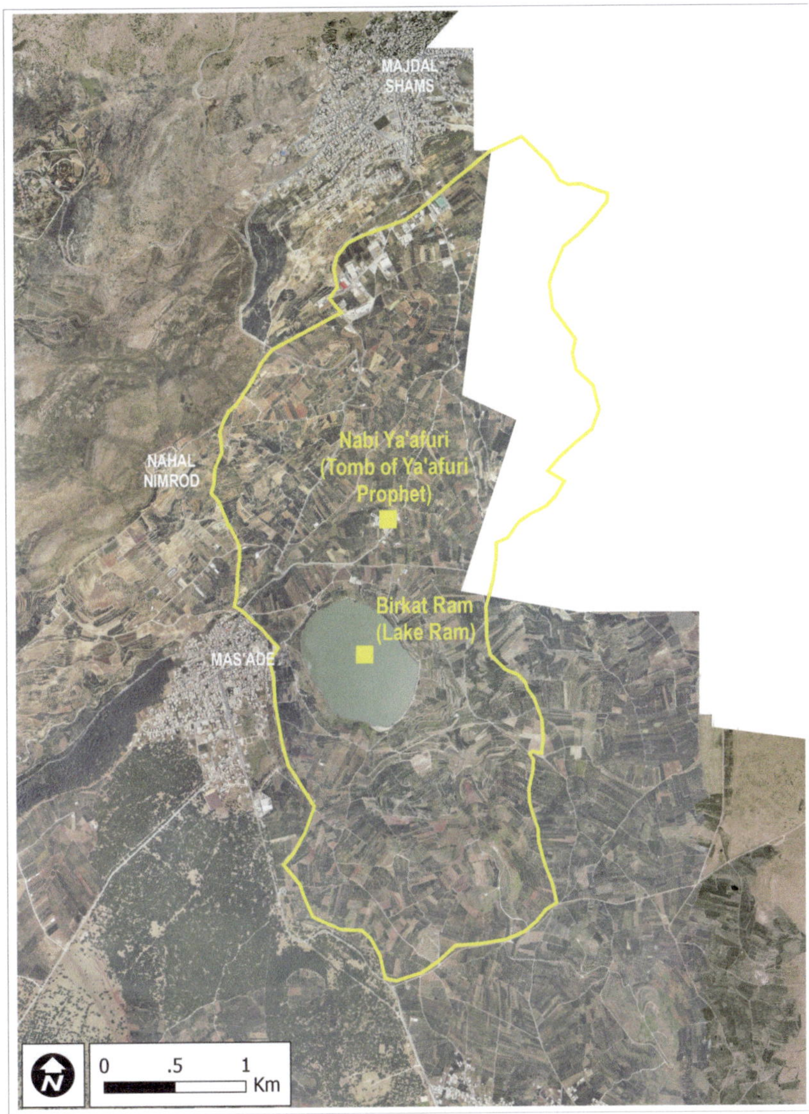

Fig. 5.134 Deciduous groves in the Ya'afuri Valley—polygon

the open landscape, along with the effort involved in such an expedition, were important components of the pilgrimage experience.

The polygon was established by defining the open spaces around the tombs necessary to preserve the essence of the experience. The polygon area contains the densest concentration of tombs in the country. The hills in the area are dotted with the domes that mark the tombs. The area contains the tombs of some of the most prominent figures of the Talmudic period. The polygon does not include the most visited tomb, that of Rabbi Shimon Bar Yohai in Meron. Since the sixteenth century, this site has been considered the holiest in the region. A separate cultural landscape polygon has been designated for this tomb and its surroundings.

Cultural Anchors

Rabbi Yehonatan Ben Uziel in Amuka, Honi Hama'agel and his wife adjacent to Hatzor HaGlilit, Rabbi Hillel the Amora and Rabbi Ishmael bar Yossi the Galilean near Dalton, Rabbi Bana'ah the Amora on the road to Safed, Rabbi Elazar HaModai and Jacob Ish Navoriah, near the synagogue of Navoriah in the Biria Forest, Rabbi Crosfodai the Amora near Kadita, Shma'ya and Avtalyon near Gush Halav, Yosef Ben Zimra near Kerem Ben Zimra.

Fig. 5.135 Deciduous groves in the Ya'afuri Valley viewed from above

Fig. 5.136 A grove in the Ya'afuri Valley with a metal rainwater pool

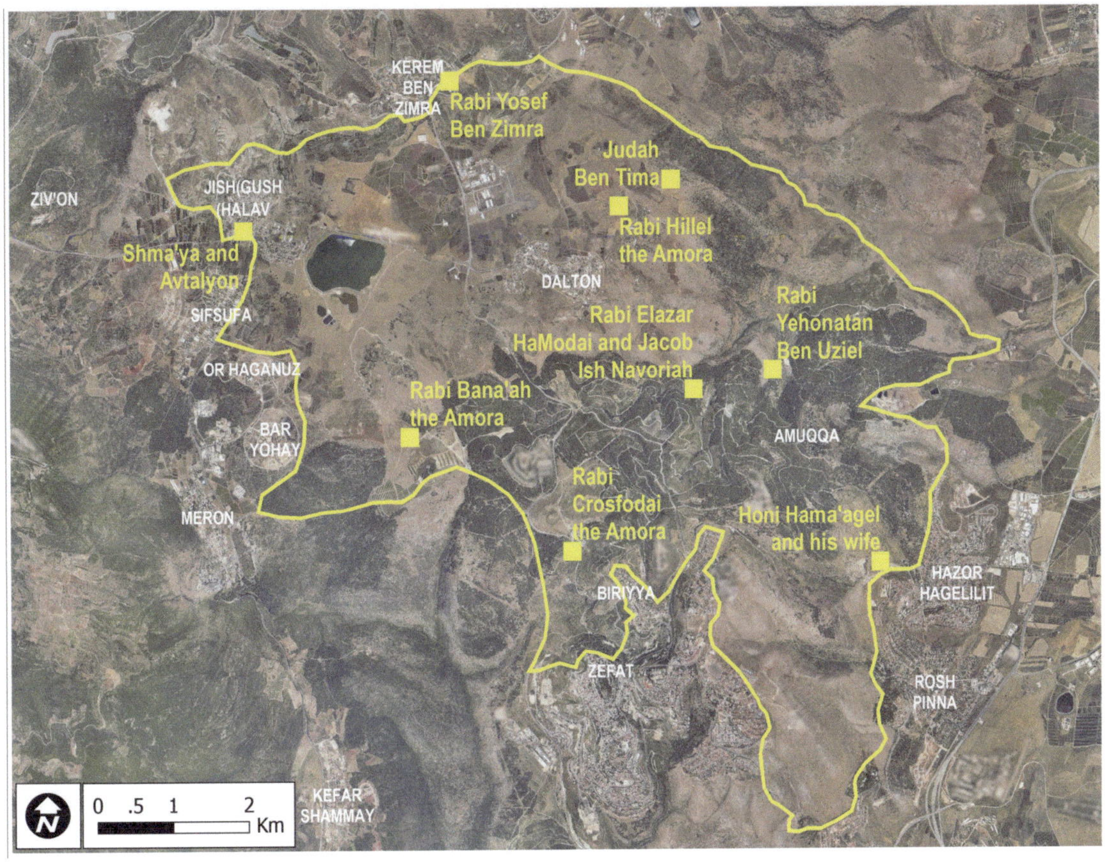

Fig. 5.137 Tombs of the Righteous (Tzadikim) in the Galilee—polygon

5.2.4.2 The Tomb of Rabbi Shimon Bar Yohai (Fig. 5.140)

Narrative

Meron was an important Jewish settlement during the Second Temple period that continued to flourish during the Roman period. It is believed that the site was abandoned during the late fourth or early fifth century CE, perhaps under taxation pressures from the authorities. One of the most impressive ruins of the ancient site is a synagogue dating from the beginning of the Byzantine period, which is among the most notable Galilean synagogues from that time. During the Muslim period, an Arab village was established on the site of the graves of Talmudic sage Rabbi Shimon Bar Yohai (Rashbi) and his son Elazar are found at the edge of the village.

Rabbi Shimon Bar Yohai was one of Rabbi Akiva's most outstanding students, who resisted the Romans and hid from them for thirteen years.[6] During his later years, he lived in Meron, and it is said that he wrote the Zohar there. For many centuries, his death has been traditionally commemorated on the holiday of Lag B'Omer.[7] This annual event draws many thousands of pilgrims.

Defining Polygons

Rabbi Shimon Bar Yohai's Tomb and the ancient synagogue are located within or adjacent to the declared nature reserve on Mount Meron. Despite the high level of statutory protection, there has been a long trend of uncontrolled construction for many years. Numerous buildings have been built to fulfill the needs of the many pilgrims visiting the site. These have severely compromised the "spirit of the place." In the vicinity of the tomb, there is a defined area where one can still feel the atmosphere of an ancient sacred place. Around the tomb complex, a buffer zone has been delineated that separates it from the settlement and the nature reserve. The buffer zone has several substandard structures that need to be removed or upgraded to reflect the character of the

[6] Initially a poor uneducated shepherd, Akiva ben Yosef became one of the greatest scholars in Jewish history.

[7] A Jewish holiday on the eighteenth day of the Hebrew month of Iyar, celebrated on the 33rd day of the Counting of the Omer, which ends a period of mourning and a number of associated restrictions.

Fig. 5.138 New construction surrounding the historic tomb

place. It also contains open space aimed at preserving the distinct character of the tomb on the slopes of Mount Meron. Aside from the tomb, the buffer zone includes the area of the ancient synagogue, which, according to tradition, was built by Rabbi Shimon Bar Yohai.

Cultural Anchors

The tomb of Rabbi Shimon Bar Yohai consists of an array of structures surrounding a courtyard that date primarily from the late Ottoman period. A large dome covers the burial hall, which, according to tradition, covers the subterranean burial cave where Rabbi Shimon Bar Yohai was buried. Ownership of the tomb and the surrounding structures was given to the Jewish community of Safed during the nineteenth century and the community has continued to maintain the tomb.

The Cave of Hillel and his Disciples—near the tomb of Rabbi Shimon Bar Yohai there is an ancient burial complex known as the "Cave of Hillel and his Disciples." This is part of a complex system of catacombs found in ancient Meron. The cave consists of an entry room that leads into a central hall typical of Jewish burial caves from the Roman period. The cave has twenty-four burial vaults covered with stone lids and twelve exposed niches.

Additional tombs of righteous men in Meron—There are many other tombs around Meron. The most renowned are the tombs of Rabbi Yochanan HaSandlar, Rabbi Ya'aba Saba, Rabbi Himnona Saba, Rabbi Yossi Ben Kisma, Shammai the Elder and his wife (located near the town of Khirbet Shamma) and his students.

The synagogue—the synagogue, which faces Jerusalem, was built on a ledge hewn into the stone cliff that rose above the ancient settlement. The main hall (measuring 27 by 13.56 meters) had three rows of columns, with eight columns in each. The western wall was made of hewn rock. The western façade contains a central opening and has survived in the best condition. Its lintel was reinforced in modern times. (According to legend, the collapse of a cracked door frame signals the arrival of the Messiah). The eastern

Fig. 5.139 A historic Tomb of the Righteous in the Galilee

entrance has been restored. The lintels and doorposts lack ornamentation, although the construction reflects considerable expense. Its simple details may be due to the Talmudic sages' directive not to overdecorate synagogues.

5.2.4.3 The Peak of Mount Carmel (Rosh HaCarmel (Figs. 5.141 and 5.142)

Narrative

The Bible describes the Carmel and the sea as a single unit, like a point that penetrates the sea or touches it: "...and it reached to Carmel westward..." (Joshua 19: 26). This dramatic encounter impressed pilgrims who immortalized this location through their drawings, creating one of the most iconic and recognized images in the country. Rosh HaCarmel and the landscape seen from the top were featured in many Haifa photographs and postcards, and the State of Israel reissued stamps in the twentieth century celebrating the landscape.

The Carmel is considered a prime overlook to the sea, with much historical evidence indicating the deep visual connection between the mountain and the sea below. "And Elijah went up to the top of Carmel ... And he said to his servant: 'Go up now, look toward the sea'..." (Kings I xviii: 42–43) During the Roman period, there were likely two lighthouses at the site. The Spanish consul donated the current lighthouse in 1928. The revolving light beam, which scanned the sea and marked the city, was the source of the name given by the consul, *Stella Maris* ("Star of the Sea").

Beginning in the Canaanite period, the Carmel peak was considered sacred, a place of solitude and refuge. According to tradition, Elijah hid in a cave in the steep slopes of the mountain that penetrated the sea. Jews, Christians, Muslims, Druze and Baha'i have identified Elijah's Cave as the largest cave in the lower part of the mountain, and there is evidence that it became a pilgrimage site in the fifteenth century. In the garden of the Carmelite Monastery, there is a statue of a foot with a dedication to Zeus Heliopolitini, God of the

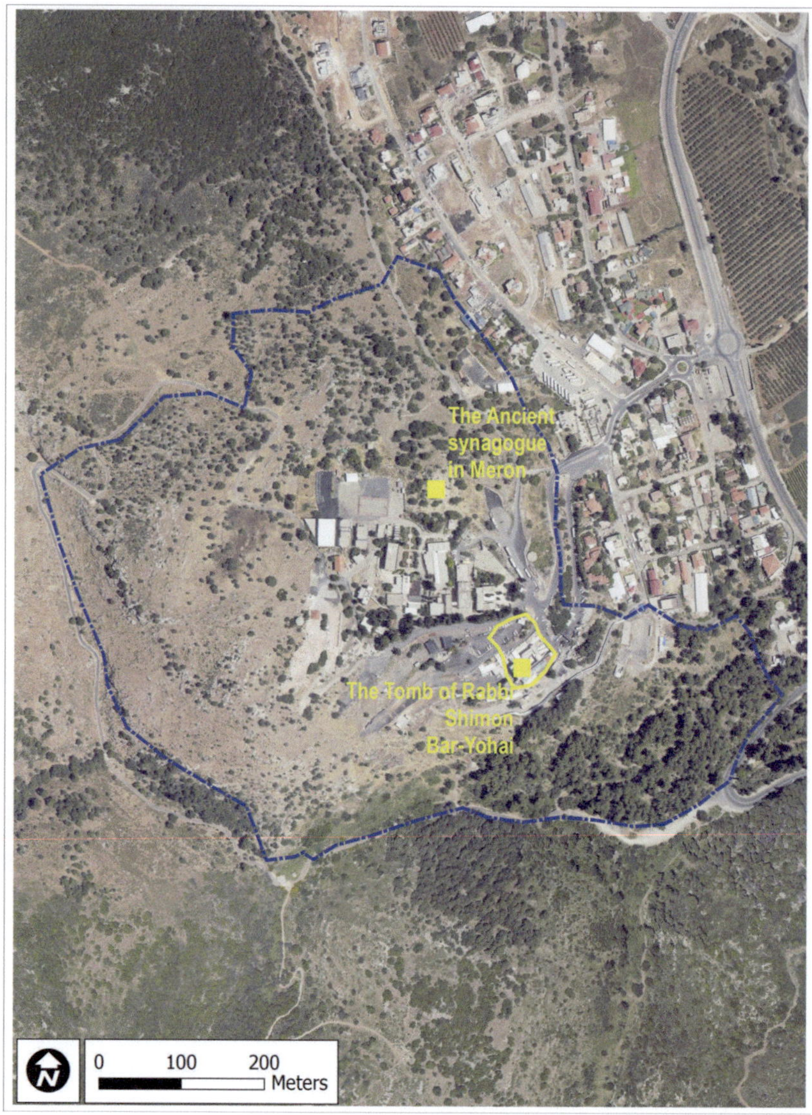

Fig. 5.140 The Tomb of Rabbi Shimon Bar Yohai—polygon

Carmel. During the Byzantine period and Crusader periods, a church existed on the site. In 1769, the French consul obtained a *firman* (an edict) from the Ottoman Sultan confirming French patronage over members of the Carmelite Order. Since then, the area has been referred to as the "French Carmel." The monastery is the world center of the Carmelite Order.

In Hebrew poetry, Rosh HaCarmel is depicted as one of the "gates to the country." For example, in Ya'akov David Kamson's 1946 poem, the peak of the Carmel, with the sea at its base, appears as a raised flag: a symbol of an eternal Israel.

The Landscape

The meeting of the Carmel Mountain and the sea is unique. Its form has evoked metaphors describing the northwestern point of the Carmel shoulder as "the peak of Carmel" (Amos 9: 3). From the top of the mountain, it is possible to see the sea from three directions: north, east, and west. The mountain plunges into the Mediterranean in a steep decline of 150 meters, leaving a narrow coastal strip of about a hundred meters. The unique waves generated by the currents at the base of the Carmel Point have made the area a unique attraction for surfers.

Fig. 5.141 The peak of Mount Carmel (Rosh HaCarmel)—polygon

From a single viewpoint, one can enjoy a panoramic view over the Haifa Bay, with the city descending the mountain towards the sea, the shoreline that continues north to Rosh Hanikra on the Lebanese border and Mount Hermon that marks the border with Syria. In the other direction, one can see all the way to Hadera and the stacks of its power plant.

Defining Polygons

Preserving the visual integrity of Rosh HaCarmel is no simple task. The development pressures are among the most intense in the country. During the past two decades, numerous infrastructure projects have been built that have undermined the natural landscape. The cable car, built in the mid-1980s links the promenade in Bat Galim to Rosh HaCarmel. The National Maritime Museum and the Clandestine Immigration and Naval Museum were both built on the narrow strip between the mountain and the sea. In addition, beneath the peak of the Carmel, is a large multi-leveled highway interchange whose proportions diminish the power of the encounter between the Carmel Mountain and the sea.

The cultural landscape polygon includes the remaining areas that have not been built up, beginning at the high ridge of the Carmelite Order, through the mountain slopes, the narrow coastal strip and sea breakwaters. Every building proposed inside the cultural landscape polygon should be

Fig. 5.142 A view of the peak of Mount Carmel, looking north toward Mount Hermon

examined in terms of its visibility from the sea and potential harm to the iconic topography that contains strong historical associations.

Cultural Anchors

The Lighthouse, Stella Maris monastery, the cave of the prophet Elijah.

5.2.4.4 "The Guard of the Fields" (Al Mishmar HaSadot)—Alexander Zaïd (Figs. 5.143, 5.144 and 5.145)

Narrative

Already during his lifetime, Alexander Zaïd was a symbol of renewed Jewish settlement in Palestine. Zaïd was born in 1886 in Russia. An intoxicated Russian farmer killed his mother when he was three, and his father, a Hebrew scholar and Zionist, spent 25 years in exile. At the age of 13, Alexander joined the Zionist Labor Movement, and in 1904 he was one of the early pioneers of the Second Aliyah.

Zaïd was one of the founders of the secret organization *Bar Giora* (1907), whose purpose was to protect Jewish pioneers. In 1909, this group spawned a new organization called *HaShomer* (The Watchman). Zaïd joined a group of members of *HaShomer* who settled in the Upper Galilee and later founded Kibbutz Kfar Giladi. In 1926, Zaïd left the group and moved to an area between the present-day Nahalal and Yagur, serving as a watchman for the Jewish National Fund. His watch included communities, planted forests and fields in the valley.

Zaïd made his home in the hills of Sheikh Abreik, and despite many hardships, he and his family were steadfast in their claim to the land. Many viewed him as a model of faith, courage and sacrifice. He pressured the settling institutions to establish additional communities in the region. Zaïd was also connected to the *Noar HaOved* youth movement and appeared before a large gathering where he encouraged them to establish a "kibbutz of shepherds," which was to become Kibbutz Alonim. His reputation as a generous host spread, as well as his efforts in cultivating rapport and good neighborly

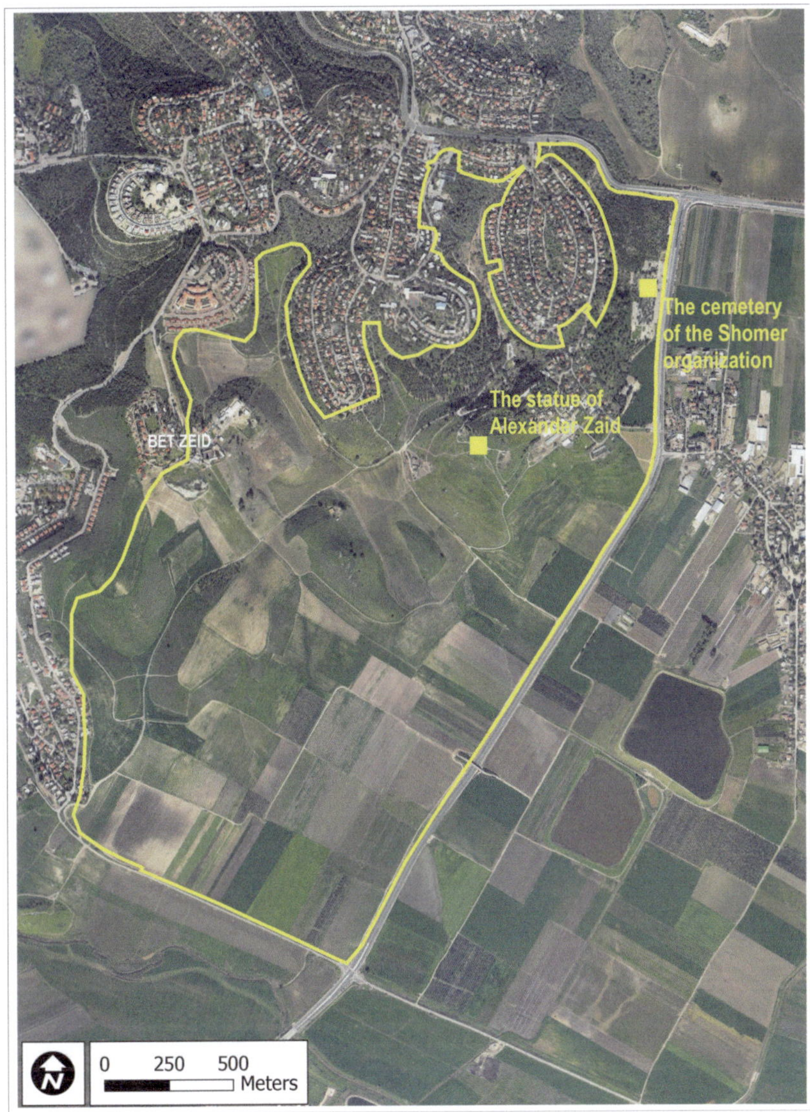

Fig. 5.143 "The Guard of the Fields" (Al Mishmar HaSadot)—Alexander Zaïd—polygon

relations with local Arabs, although ultimately, these efforts did not bear fruit. On July 10, 1938, amid the "Great Arab Revolt," as the bell rang to call the kibbutz members to a meeting, he went out for the last time. He was shot and killed by Qassem Tabash, a Bedouin assassin. Zaïd's funeral drew a massive crowd. The leaders of the Yishuv spoke, establishing Zaïd's place in the founding stories of the nascent state.

The Landscape

In 1940, sculptor David Polus, an immigrant from the Third Aliyah, created an equestrian statue of Zaïd riding a horse and placed it near the Sheikh Abreik tomb and the Zaïd family home. The Zaïd statue faces east, gazing over the Jezreel Valley. The small statue became a landmark in the landscape of the Jezreel Valley and was a destination for visitors in the 1940s and 1950s. The original statue, made of wire and concrete, was replaced in 1979 with a bronze statue by sculptor Dan Zaritzky. It symbolizes the continuity of settlement in the Jezreel Valley and its security.

The site of the renowned watchman's murder is located in what is now the town of Kiryat Tivon and has been preserved as a public garden. Nearby, shaded by tall pines, is the cemetery of the *Shomer* (The Watchman) Association. Alexander Zaïd and his wife Zipporah are buried there, as is the writer Eliezer Smoli, author of "Lives of the Founders," the story of the Zaïd family.

Fig. 5.144 The view toward Sheikh Abreik

Defining Polygons

The cultural landscape of fields and forests where Alexander Zaïd stood guard includes the surviving agricultural landscape and the reforested areas planted by the Jewish National Fund. The landscape of agricultural fields interspersed with hills of oak trees has largely been preserved around the Zaïd house, which stands next to the famous statue.

Some of the original fields and forests have given way to the development of the modern town of Kiryat Tivon. Kibbutz Alonim relocated to a new site. Kiryat Amal is located on the original site of the "shepherds' kibbutz" in the forest. The town of Kiryat Tivon is built on the slopes of hills that preserve the local Tabor oak trees indigenous to the region. In the valleys between the built-up hills, there are open spaces where crops are grown, primarily wheat. The landscape of the Alexander Zaïd story is near the landscape of Elijah and the prophets of Ba'al story (see Sect. 5.2.4.7). National infrastructure projects slated for implementation are projected to completely change the little that has survived in the unique intersection between the open fields and natural oak forest on the hills of Sheikh Abreik.

Cultural Anchors

Statue of Alexander Zaïd in Sheikh Abreik, the cemetery of the Kiryat Tivon HaShomer organization.

5.2.4.5 King Saul and the Philistines at the Gilboa (Figs. 5.146 and 5.147)

Narrative

The Gilboa Mountains and the area at its base witnessed a tragic battle between the Philistines and the Israelites, which culminated in the death of King Saul and his sons. Saul set his camp in Ein Jezreel, on the western slopes of Mount Gilboa. When he saw the might of the Philistine army, he

Fig. 5.145 Alexander Zaïd—equestrian statue by David Polus

became fearful. Saul asked God for advice, and when he did not receive an answer, he sought advice from the Witch of Endor. The king disguised himself and went to the witch at night, where she told him that the Philistines would overpower him: "… Moreover, the Lord will deliver Israel also with thee into the hand of the Philistines; and tomorrow shalt thou and thy sons be with me; the Lord will deliver the host of Israel also into the hand of the Philistines" (I Samuel 28: 19). The witch's words came to pass. The children of Israel fled to Mount Gilboa, with the Philistines inflicting many casualties, killing Jonathan, Avinadab and Malkishua. Saul was afraid to be captured alive, lest he be tortured. He asked his arms-bearer to kill him but was refused. Saul decided to take his own life: "… and Saul took his sword and fell on it" (I Samuel 31: 5).

The Philistines found the bodies of the king and his sons and hung the corpses on the wall of Beit She'an. The people of Jabesh-Gilead in Trans-Jordan came to save the honor of the king, who had protected them against Nahash of Ammon. A group of men from Jabesh Gilead crossed the Jordan River at night and came to Beit She'an; they removed the bodies from the wall and brought them to Jabesh Gilead for burial: "And they took their bones and buried them under the tamarisk-tree in Jabesh, and fasted seven days" (I Samuel 31: 13).

The Landscape

The top of the Moreh Hill and Mount Gilboa offer panoramic views of the Harod Valley and the Beit She'an Valley, where Saul and his sons fought their final battle. At the northern base of the Moreh hills is Khirbet Eindor, identified as the biblical site Ein Dor. One kilometer to the northeast is Khirbet Tzfafot, the spring of the biblical village. From the hill, it is also possible to see Saul's final route from Ein Dor back to the camp in Ein Jezreel. There are ten kilometers between the Ein Dor and Ein Jezreel. Saul must have crossed the Harod stream on his journey. We can assume that even then, as today, local residents took advantage of the stream's water to irrigate the nearby slopes. From the peaks of the Gilboa, it was possible to see the walls of Beit She'an. The biblical *tel* of Beit She'an is clearly visible from the top of Mount Barkan and Mount Avinadav. From here, one can see Jabesh Gilead (near Wadi Yavesh) in Jordan, where the people of Jabesh Gilead buried Saul and his sons.

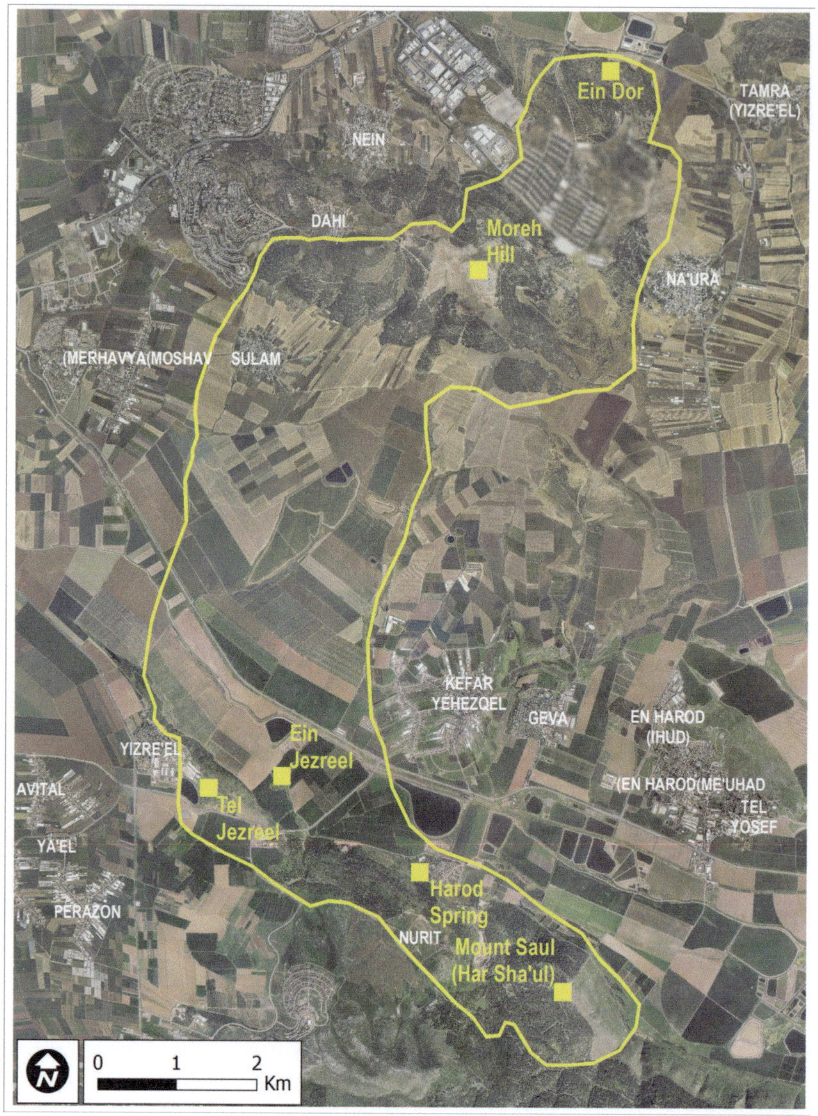

Fig. 5.146 King Saul and the Philistines at the Gilboa—polygon

Cultural Anchors

Moreh Hill, Ein Dor, Ein Jezreel, Tel Jezreel, Mount Saul (Har Shaul), Harod Spring.

5.2.4.6 Gate of the Valley (Sha'ar Hagai) (Figs. 5.148, 5.149 and 5.150)

Narrative

Until the second half of the nineteenth century, the main road linking Jaffa to Jerusalem crossed the Beit Horon ascent. While this route was longer than the road that crosses Sha'ar Hagai, it avoided the difficult descent to the Kisalon stream and the steep ascent near the Castel. In the second half of the nineteenth century, the number of pilgrims to Jerusalem increased significantly. The Ottoman authorities decided to build a road for carriages along the bed of the Nahshon stream. The road construction through Sha'ar Hagai was completed in 1869, and this has been the main road to Jerusalem ever since.

During the 1948 War of Independence, fierce battles took place along Sha'ar Hagai for eight months. The road, winding for six kilometers through the valley, was easily ambushed. On the ridge overlooking the valley from the south were the Arab villages of Beit Machsir and Saris, while the northern ridge had posts occupied by Haj Amin Al Husseini's armed men. Jewish forces tried to break the siege of Jerusalem in convoys of supply trucks, with their cabs armored with protective steel plates. Although some command points were conquered, such as the Castel and the Masrek, the road remained blocked to Jewish passage.

The Landscape

Sha'ar Hagai, (*Bab el-Wad* in Arabic) is the name of the "natural gateway" in the mountains, which was created in

Fig. 5.147 King Saul and the Philistines at the Gilboa

the places where Nahshon stream emerges from the harsh limestone ridges of the Judean Mountains as it shifts to the softer foothills. Over the past one hundred and fifty years, this section of the road to Jerusalem has served as the city's western gate. Until the establishment of the State of Israel, the slopes were primarily covered with ancient agricultural terraces. Forests were planted along the road during the British mandatory period, and this afforestation process accelerated with the establishment of the State of Israel. Monoculture plantings of Aleppo pine trees *(Pinus halepensis)* by the Jewish National Fund have transformed the trip through the valley into a journey through a thickly forested landscape. In recent years, the forest has suffered from diseases and forest fires. The Jewish National Fund is currently diversifying the forest on the slopes, planting local woodland species and fruit trees.

The roadside contains the ruins of the armored vehicles damaged while attempting to break the siege. There were several initiatives to remove them, but the public protested. The road became a memorial site commemorating the fierce battles of the War of Independence in general and the breaking of the siege of Jerusalem in particular. Since then, the road has been upgraded several times, and there is almost no trace of the original narrow road along the riverbed. Each time the road was expanded, the remains of the armored vehicles were transferred. On Independence Day, many visitors come to pay their respects at commemoration ceremonies.

Defining Polygons

The Sha'ar Hagai area is well protected as part of the National Master Plan 8 for Parks and Nature Reserves. It is also included in National Master Plan 21 for the War of Independence. Sha'ar Hagai is an important part of Rabin Park, and is a bridge between the section of park in the Judean Hills and the section in the Judean Mountains. The

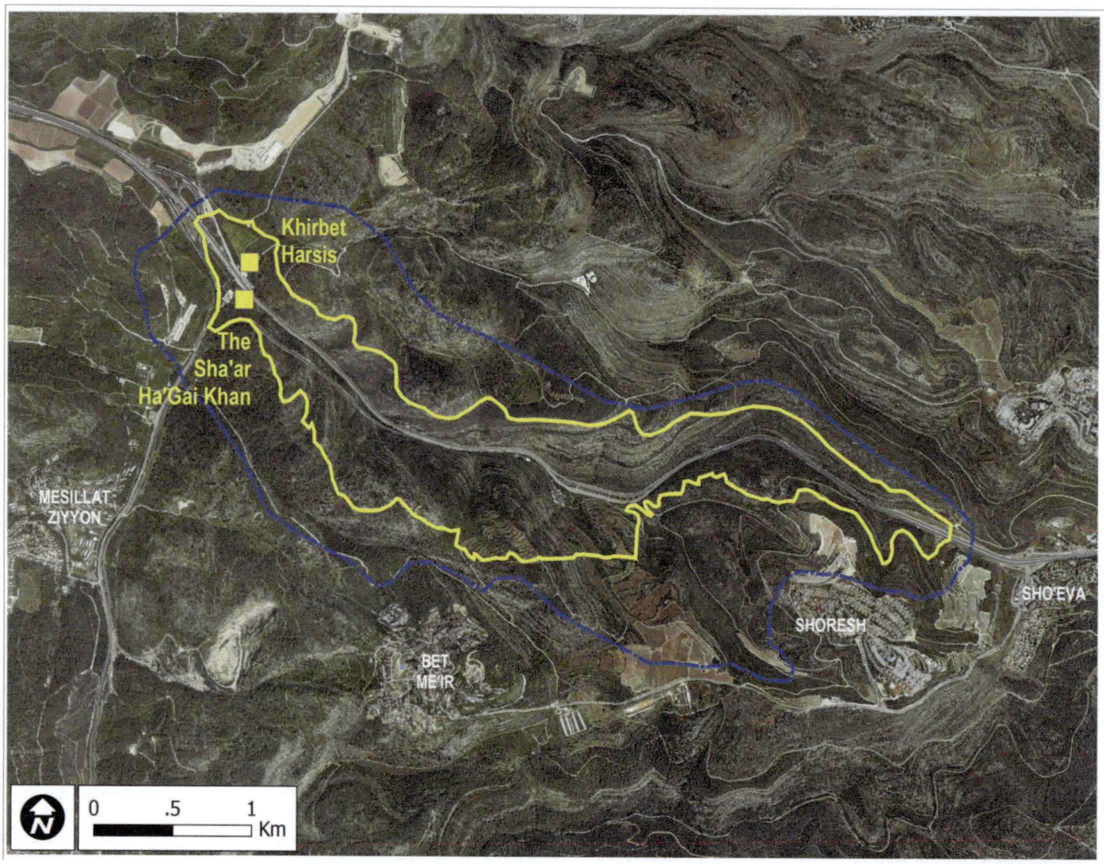

Fig. 5.148 Gate of the Valley (Sha'ar Hagai)—polygon

Fig. 5.149 The ruins of armored vehicles displayed at Sha'ar Hagai

Fig. 5.150 Sha'ar Hagai

cultural landscape polygon includes the Nahshon stream, the ridge of outposts to the north, and the ridge of convoys to the south, commemorating the supply convoys to Jerusalem.

Cultural Anchors

There are two buildings at the entrance to Sha'ar Hagai:

Khirbet Harsis—to the north of the road is one of 17 watch stations along the route. Each building had a group of horsemen whose job was to protect their section road from thieves.

Khan Sha'ar Hagai—south of the road is a *khan* (inn) and a horse changing station that served travelers. Renovation work on the *khan* began in 1997, and there are plans to turn it into a visitors' center.

5.2.4.7 Elijah and the Prophets of Ba'al (Figs. 5.151, 5.152 and 5.153)

Narrative

The biblical book of Kings (XVIII 19–40) describes one of the most famous battles of monotheism against idolatry: the struggle between Elijah and the prophets of Ba'al, between the sole representative of a supreme God and hundreds of pagan representatives. In a competition held on Mount Carmel, Elijah proved that he was a true prophet when God accepted his plea and sent fire from the sky to consume his burnt offering. The prophets of Ba'al failed. A furious Elijah took advantage of the success of the miracle, rallied the people and pursued the prophets of Ba'al down the slopes of Mount Carmel to the Kishon River, where he slaughtered four hundred Ba'al followers. When Jezebel, the Sidonian wife of Ahab, King of Israel, heard the news, she threatened to execute Elijah, and he fled to the desert.

The Landscape

The Carmel Horn is a steep and isolated mountain towering above the Jezreel Valley, surrounded by tangled Mediterranean scrub, making it an extremely suitable place for worship. In the nearby Khirbat Kalia, evidence indicated that this was likely an ancient ritual site, later given the Hebrew name *Metsudat Habama*.

The tradition that identifies Mount Carmel as the scene of the struggle between Elijah and the prophets of Ba'al is quite ancient. During the fourteenth century, an anonymous disciple of the Catalan rabbi and philosopher, Nachmanides, visited the site and saw Elijah's altar. In Arabic, the site is

Fig. 5.151 Elijah and the prophets of Ba'al—polygon

called *Al-Muharraqa*, derived from the Arabic word for charring and fire, referencing the fire that came down from heaven to consume Elijah's sacrifice. Carmelite monks built a monastery on the site and named it after the prophet Elijah. In the monastery's courtyard, there is a statue depicting Elijah wielding a sword with one of the prophets of Ba'al at his feet. Three reliefs on the walls of the courtyard describe the biblical narrative. The Carmelite site is well suited to the biblical story of the slaying of the prophets of Ba'al. We can assume that those who tried to flee had difficulty escaping through the scrub, enabling their pursuers to capture and bring them to the banks of the Kishon River.

Many pilgrims flock to the site to enjoy the magnificent panoramic views of the Carmel and the Jezreel Valley from the monastery's roof. It is possible to track the geographical progression of the biblical story, from the altar atop the Carmel to the flight of the prophets of Ba'al to the Kishon.

Defining Polygons

The polygon includes the ancient site of worship, Metsudat Habamah, the Carmelite Monastery on the Muharraqa, and the view from the slopes of the Carmel to the banks of the Kishon. The Carmel in this region still resembles the biblical landscape of "Kerem El," a wild area unmarred by agriculture and development. The Kishon River around Tel Kashish primarily reflects the natural landscape of the river, and the open space around it, the valley where the prophets of Ba'al were slaughtered.

The area around Keren HaCarmel is protected, so it is impossible to build there. As a result, the extensive overgrowth of trees makes it challenging to find an observation point to seat a group to tell the biblical story. The best available viewing position is from the roof of the monastery. It is particularly difficult to find a vantage point to the west overlooking the Mediterranean Sea, the direction of the rain cloud which rained on the altars of Elijah and the prophets of Ba'al. Thus, preserving the natural vegetation restricts the visitor's view and one's ability to experience Mount Carmel associatively.

Cultural Anchors

Keren Carmel (the Carmel Horn), Metsudat Habamah (Khirbet Kalia).

Fig. 5.152 Elijah and the prophets of Ba'al

Fig. 5.153 The Carmelite monastery at *Al-Muharraqa* from where Elijah pursued the prophets of Ba'al

5.2.4.8 David and Goliath (Figs. 5.154 and 5.155)

Narrative

The story of David and Goliath appears in the book of Samuel, Chapter XVII. Biblical scholars date the text to the end of the second millennium BCE. The story tells of the victory of the Israelite boy, David, in the fight against the Philistine giant, Goliath. This is one of the most dramatic and famous narratives in the Bible. The story of this event shaped the image of the Kingdom of Judah for 400 years, becoming one of the most widely known stories in the Jewish tradition. It became the prototype for all conflicts where the weak overcome the strong because of their righteousness and wisdom.

The battle between David and Goliath took place in the Elah Valley. The valley originates in the hills of Bethlehem, the birthplace of King David, and is named for its large terebinth trees (*elah* in Hebrew). From the hills of Bethlehem, the valley winds through the mountains, across the land of the Philistines, to the Mediterranean. Due to its location and geographic structure, the valley became one of the most important crossroads between the Judean Hills and Jerusalem, Hebron and Bethlehem. Several cities surrounded the valley at the time of the battle, including Azekah, Sochoh and Adullam.

The battle began by establishing two camps in the region. The Philistine camp was situated in Sochoh, in the southern part of the valley; the Israelites camped in Azekah, a town overlooking the valley from the west. The Elah stream flowed between the two camps. Each side was entrenched on a separate hillside. The valley, encircled by the two well-fortified camps, became the stage set for the theater of war. After David refused to use conventional arms, Goliath and David ran toward each other. David used a slingshot, hitting Goliath in the head with a stone. Goliath collapsed, and David cut off his head with the vanquished giant's own sword. After watching the shameful defeat, the Philistines fled west, down the slope of the Elah stream, with the Israelite army at their heels, inflicting many casualties along the way. The escape route led the fleeing army to the fortified Philistine cities of Gat and Ekron. The location of Ekron has been identified as Tel Makena (Khirbet al-Mokna) near Kibbutz Revadim. The location of the city of Gat is associated with Tel Safit (Tel el-Safi) near Kibbutz Kfar Menachem. The battleground known as *Efes Damim* was recently identified as Khirbet Qeiyafa. Since this identification was made, and due to the great historical significance of the David and Goliath story, in 2015, the Ministry of Interior and the Ministry of Housing decided to forgo the construction of 1600 housing units approved in the city of Beit Shemesh.

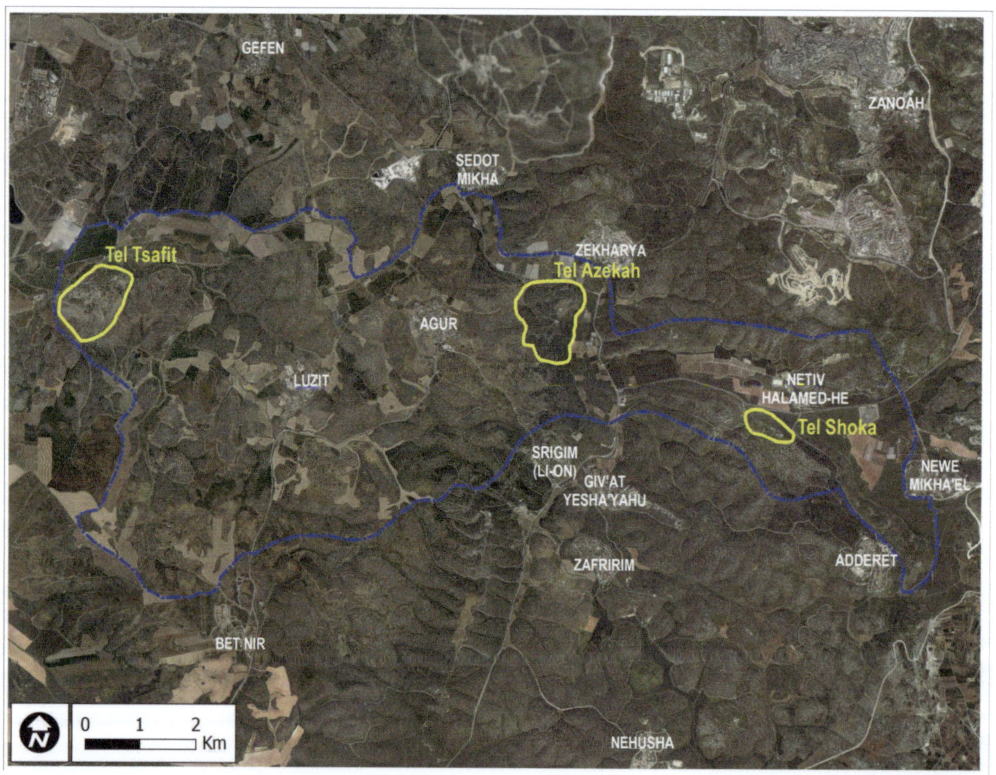

Fig. 5.154 David and Goliath—polygon

Fig. 5.155 David and Goliath—view of the Elah Valley

The Landscape

The Elah stream descends from the Hebron Mountains and flows north between the mountain and the Judean Lowlands towards the foothills of Jerusalem, where it turns west. From that turning point, the river crosses a ridge of hills in the Judean Hills to the broad and beautiful Elah Valley—a level area bordered by an area of steep slopes. This is an "eroded corridor" typical of rivers crossing the Judean Lowlands.

Defining Polygons

Today the Elah Valley is settled mainly by farming communities, including Kibbutz Netiv Halamed Hey, Givat Yeshayahu, Tzafririm, Roglot and Aviezer. Currently, the communities have little visual impact, and it is possible to imagine the area as it was during the battle between David and Goliath. The limits of the polygon include the entire valley and mountains that frame it. The polygon includes the region of Tel Azekah to the west until the city of Gat (Tel Tzafit), where the Philistine fighters fled after David slew Goliath.

Cultural Anchors

The points where the story is best illustrated are the three *tels* identified in the cultural landscape.

Tel Shoka—the Philistine army camped at the base of the *tel*. From the top, it is possible to see Bethlehem and the battlefield of the Elah Valley.

Tel Azekah—the best vantage point for understanding the entire story is from the top of Tel Azekah. One can look out over the Elah Valley, the row of *tels* (that include Tel Azekah). It is also possible to see the Hebron Hills from where the Israelites approached and the Judean Hills, the direction the Philistines fled.

Tel Tzafit—at the top of the *tel,* it is possible to see the Mediterranean Sea to the west and the coastal Philistine cities Ashdod and Ashkelon. To the east are the Hebron Hills. The agricultural landscape at the foot of the *tel* is a superb example of the landscape of the Judean Hills, which is distinct from the landscapes at the foot of Tel Shoka or Tel Azekah.

Tel Qeiyafa—(Hirbet Kaifeh) was also identified as the city of Sha'arayim, located between Tel Shoka and Tel Azekah on the northern ridge of the Elah Valley. A wall surrounding the city enclosed the site, and evidence was found linking the site to the Iron Age, the era of King David.

References

Dio Cassius, The *History of Rome*, Volume 8, Book 69, paragraph 12.

Epstein, C. (1974). The Dolmens in the Golan, *Qadmoniot, 7*(2–1), 37–40. (Hebrew).

Pliny the Elder, *Historia Naturalis*, Pliny the Elder, Book 12: XXXII).

Vinitzky, L. (1991). The Age of the dolmens in the Golan and the Galilee: A new review. *Eretz Yisrael, 21*, 167–173. (Hebrew).

6 Planning and Management

As landscapes are inherently dynamic, the protection of cultural landscapes depends upon a management framework to guide change in the landscape while retaining important values "for present and future generations" (UNESCO, 2021, Paragraph 109). The Operational Guidelines state that "management systems may vary according to different cultural perspectives, the resources available and other factors. They may incorporate traditional practices, existing urban or regional planning instruments, and other planning control mechanisms, both formal and informal" (UNESCO, 2021, Paragraph 110). It is critical to reach "a thorough shared understanding of the property" and its values by all stakeholders, including local communities and indigenous peoples (UNESCO, 2021, Paragraph 111). The Operational Guidelines stress the need for developing mechanisms for inclusive and participatory planning processes and monitoring and feedback cycles over time. Capacity building, sufficient resource allocation, and transparency are additional factors for successful management (UNESCO, 2021, Paragraph 111).

The Operational Guidelines also address the need for "short, medium and long-term actions to protect, conserve and present the nominated property" and state that "an integrated approach to planning and management is essential to guide the evolution of properties over time and to ensure maintenance of all aspects of their Outstanding Universal Value" (UNESCO, 2021, Paragraph 112). The goal is to go beyond the boundaries of the properties and address the wider landscape setting.[1]

The Israel Nature and Parks Authority (INPA) was tasked with enforcing the National Parks, Nature Reserves, National Sites, and Memorial Sites Law (1998) and managing the National Parks and Nature Reserves. According to the law, "The functions of the Authority are to deal with all matters pertaining to the nature reserves and national parks and to promote their interests, and also to protect the natural assets and heritage, to supervise their preservation and to enhance them, including: … to establish, manage, maintain, operate and enhance nature reserves and national parks…".[2]

This chapter presents three important management case studies developed in Israel over the last two decades. Two sites were deemed of international importance and have been nominated and inscribed within the UNESCO frameworks, World Heritage Convention and the Man and Biosphere Program. A third plan was approved as a partial District Outline Plan with a focus on participatory planning. These case studies describe the processes that enabled these cultural landscapes to be protected at the national and international levels, as well as the interrelationships among the different levels of planning. We focus on the stakeholders, the public participation mechanisms, and the translation of the broader cultural landscape terms into planning actions and implementation.

The first plan discussed is the Incense Route—Desert Cities in the Negev, inscribed as a World Heritage Site in 2005 (see Chap. 5). The second is the Megiddo Biosphere Region (formerly known as Ramot Menashe Biosphere Region), a Man and Biosphere Reserve declared in 2010 and renamed in 2017.[3] The third is a partial District Plan for

[1] See UNESCO 2021, Operational Guidelines, Paragraph 112: "the wider setting may relate to the property's topography, natural and built environment, and other elements such as infrastructure, land use patterns, spatial organization, and visual relationships. It may also include related social and cultural practices, economic processes, and other intangible dimensions of heritage such as perceptions and associations… Its effective management may also contribute to sustainable development, through harnessing the reciprocal benefits for heritage and society."

[2] See Paragraph. 4 of the law: https://www.gov.il/BlobFolder/guide/natl_parks_nature_reserves_council/en/open_area_natl_parks__nature_reserves_natl_sites_and_memorial_sites_law_1998_eng.pdf

[3] https://unesdoc.unesco.org/query?q=Conference:%20%22International%20Coordinating%20Council%20of%20the%20Programme%20on%20Man%20and%20the%20Biosphere,%2029th,%20Paris,%202017%22&sf=sf:*

A. Sar Shalom et al., *Cultural Landscapes of Israel*,
https://doi.org/10.1007/978-3-031-33685-0_6

Table 6.1 Areas of inscribed property and buffer zones

Component parts	Area of inscribed property in hectares	Area of buffer zone in hectares
The route and Avdat	6 ,314	62,592
Haluza	52	278
Mamshit	242	514
Shivta	47	484
Total	6,655	63,868

Shikma Park. Each case study includes an outline of the plan and its management elements and an overview of the involved stakeholders.

6.1 The Incense Route—Desert Cities in the Negev

The UNESCO World Heritage nomination included a management plan for the Incense Route (2005). It is a linear property with a route over 65 km long and approximately 1 km wide and managed by the Israel Nature and Parks Authority as a national park.

6.1.1 The Plan

"The management system and plan are divided into the Incense Route Park and the ancient towns along the route or towns that developed as part of the route's historical role and context. The park includes fortresses, road segments, water installations, and remains of worship sites and agricultural plants" (Incense Route UNESCO Nomination, 2005, p. 120).

As a serial inscription, the four nominated cities are separate components with common historical, cultural, and social values, including extensive economic benefits (Table 6.1).

6.1.2 Plan Elements

The management plan includes a comprehensive section on roads and access, ecotourism, and recreation, such as campsites, trails, visitor centers in Avdat and Mamsheet. Archaeological sites and routes in the designated area are slated for comprehensive conservation. The nature conservation component of the plan includes the proposed reintroduction of endangered species, such as the wild ass and Arabian oryx, as well as the control of invasive species. The rehabilitation of disturbed habitats and quarries is also proposed, along with additional research and monitoring activities. Educational and professional staff training on conservation and public participation is another significant aspect of the management plan.

6.1.3 Stakeholders

The primary stakeholders of the Incense Route are the local communities and residents, mainly Jews and Bedouins, the Ramat Negev Regional Council, the Middle Arava Regional Council and the Mitzpe Ramon Local Council. Regional stakeholders include the Negev Development Authority, while national and international environmental NGOs, including the Society for the Protection of Nature in Israel and EcoPeace, have also provided support. Within Israel, other stakeholders include Ben Gurion University, which has researched ancient agriculture, the Israel Antiquities Authority, and the Israel Nature and Parks Authority (INPA), which is responsible for managing and protecting the Desert Cities and the Incense Route. Three government ministries are also involved (the Ministry of Education, the Ministry for Environmental Protection, and the Ministry of Tourism). As the site was inscribed on the World Heritage List in 2005, the World Heritage Committee is also an important stakeholder in defining the criteria and management standards required to retain the site's international status.

In 2016, the property, including the buffer zone, was included in the Beersheba Metropolitan Plan and was defined in the amendment of the preservation and development outline plan in National Outline Plan 35 as a "landscape ensemble" (cultural landscape) using the boundaries established in this volume. The inclusion in the National Outline Plan provides an additional layer of protection to ensure the site remains safeguarded over time.

6.2 The Megiddo Biosphere Reserve (formerly the Ramot Menashe Biosphere Reserve)

The UNESCO Biosphere Reserve concept emerged in 1971, narrowly defined initially as a research and conservation program for single sites with high ecological value.[4] Over the following decades, the Biosphere Reserve model evolved

[4] Biosphere Reserves were based on the Man and the Biosphere Programme (MAB), established in 1971.

as conservation became more broadly understood, absorbing the social and economic aspects central to the concept of sustainability. In 1995, the Biosphere Reserve was redefined as a more regionally oriented, ecosystem-based model emphasizing people-centered conservation and sustainable development (Coetzer et al., 2014).[5] In addition to its initial focus on ecosystem conservation and research, the Reserve later included sustainable development as a central principle (UNESCO, 1996). To balance conservation and development goals, the Biosphere Reserve was separated into three physical zones with varying levels of protection. The *core* was devoted to conservation and restricted to human use, while activities compatible with environmental protection, such as ecotourism or recreation, were permitted in the adjacent *buffer zone*. More intensive land uses were permitted in the surrounding *transition zone*, such as human settlements or agricultural areas (Coetzer et al., 2014; UNESCO, 1995).

The Megiddo Biosphere Reserve region has high cultural and natural significance. Located in northern Israel, the Biosphere Reserve is adjacent to the Mount Carmel Biosphere Reserve to the northwest. It includes Tel Megiddo, an inscribed World Heritage site (see Megiddo and Armageddon, an Associative Landscape, Chap. 5). Continuously occupied since the Chalcolithic period, the region contains important prehistoric archaeological sites. From an ecological perspective, the Megiddo Biosphere Reserve comprises several Mediterranean ecosystems. The region was settled over the past 10,000 years by different civilizations. Its early woodlands, composed of Mount Tabor oak (*Quercus ithaburensis*) and Atlantic terebinth (*Pistacia atlantica*), have been significantly reduced in the last decades. Many slopes have undergone afforestation and are currently covered with pine trees. Moderate slopes have been planted with perennial orchards, while the valleys contain annual crops. This semiarid dryland contains several small springs that feed perennial streams, increasing the region's biodiversity (Ramot Menashe Biosphere Reserve Nomination, 2010).

The Megiddo Biosphere Reserve is located in the Mediterranean biome, which makes up approximately one percent of the world's land mass. Growing populations and expanding sprawl are a particular threat to regions in this biome. It is described in the following way in the nomination: "This hilly natural/rural landscape comprises a mosaic of woodlands, shrublands, and scrub grasslands, which are either protected or are under-regulated grazing and cultivation regimes due to their location at the eastern end of the Mediterranean Basin and within the mid-section of the aridity-gradient/ecotone linking the Sahara-Arabian climatic deserts with the Asia-minor and the northern Mediterranean mesic climates. The proposed Biosphere Reserve harbors a unique community jointly assembled by Mediterranean, Asian and European-origin biotas" (Ramot Menashe Biosphere Reserve Nomination, 2010).

Over the past century, there has been a shift in the land use in the Biosphere Reserve. At the beginning of the last century, the area had both nomadic and sedentary communities with grazing animals and crops dependent on rainfall cycles. Today agriculture is irrigated, and grazing is limited to specific areas. In the second half of the twentieth century, there was an increase in employment unrelated to agriculture, though the area still retains a rural atmosphere.

Within the Biosphere Reserve, more than a dozen cooperative agricultural communities were founded in the first half of the twentieth century. These include both kibbutzim and moshavim. Jewish immigrants began to settle the area in 1922 after the Jewish National Fund purchased the land from Arab landowners. "This community was involved in the emergence of the evolving cooperative community system of the Israeli kibbutz and moshav. This system was instrumental in the success of Israeli agriculture, constituting a cornerstone in the economic base of the young State of Israel. In recent decades the cooperative kibbutz and moshav systems have undergone radical changes, expressed in economic and societal transformation and associated with the degradation of social and cultural values adopted by the original cooperative movement. However, the communities of the Biosphere Reserve are dedicated to preserving norms inherent to the cooperative movement and the moral and cultural values it attributed to agricultural work and rural livelihoods" (Ramot Menashe Biosphere Reserve Nomination, 2010).

6.2.1 Plan

The Biosphere Reserve provides an example of an area used by a dryland rural community whose economic base is agriculture. The ecologically sustainable grassland and scrubland (*batha*) maintain the forage-provision ecosystem, the base for the widely practiced dryland livestock husbandry livelihood in the Biosphere Reserve, as seen throughout the Middle East. The regulated and controlled grazing by livestock maintains the areas as a scrubland (*batha*) ecosystem, which typically harbors high levels of biodiversity. This mutually beneficial equilibrium creates a sustainable practice that can be adopted by rural communities in the Mediterranean Basin and the Middle East regional drylands.

The conservation of the biodiversity assets in the Biosphere Reserve is protected by a two-pronged approach:

[5] The Seville Conference (1995) redefined the concept of the Biosphere Reserve. See UNESCO, The *Statutory Framework of the World Network of Biosphere Reserves*, 1996.

statutory protection and the cooperation of the local community living in the Biosphere Reserve, which has committed to a civil agreement. This agreement is an example of a local government acting beyond its regulatory capacity to achieve economic and ecological sustainability through the voluntary adoption of an agreed-upon environmental code of conduct. This agreement was the culmination of a complex and intensive bottom-up social and democratic process (Ramot Menashe Biosphere Reserve Nomination, 2010).

6.2.2 Plan Elements

The approved Biosphere Reserve comprises four core areas surrounded by extensive buffer zones and corridors that provide linkages across the Reserve. The currently accepted core areas were significantly reduced from the original 2008 proposals, which were amended according to UNESCO recommendations. "The core zone comprises 16.8 km^2, some ten percent of the total Biosphere Reserve area. These core areas of *batha* grassland and scrubland ecosystems are located in the southwestern part of the Biosphere Reserve. They are divided into four closely linked areas, separated by buffer zones. Approximately 35% of the core area is under the legally binding, statutory land use regulations of the Partial Northern District Master Plan 12,635/c for Ramot Menashe Park, a plan banning all human activities other than limited grazing. At the time of the nomination, just 2% of the core area was designated as a Nature Reserve, while 74% was at the beginning of the designation process" (Ramot Menashe Biosphere Reserve Nomination, 2010).

6.2.3 Stakeholders

One of the plan's unique elements is an organizational structure that created a broad range of local involvement and stakeholder participation. In 2008, the Megiddo Regional Council, national government agencies, active residents, and business community members initiated a public planning process to establish a Biosphere Region within the boundaries of the Megiddo Regional Council. The Biosphere Region concept focuses on human–environment linkages and ensures sustainable development to bequeath an ecologically sound environment for future generations. "The participatory planning process included public meetings to brainstorm and build consensus; establishing thematic work teams for formulating policies in demographic growth, agriculture, woodlands, and grasslands uses leading to sustainability; interfacing with national agencies active in the area of the proposed Biosphere Reserve for adapting their actions to promote rather than impede sustainability; engaging in public and country-wide outreach through the electronic media, and defining the proposed Biosphere Region's zonation. This planning process culminated with the ratification of the Ramot Menashe Biosphere Region Agreement in an open public meeting with all stakeholders present" (Ramat Menashe Biosphere Reserve Nomination, 2010).

There are many stakeholders actively involved in supporting the Megiddo Biosphere Reserve. The Israel Nature and Parks Authority (INPA) provides monitoring, surveying, environmental education and enforcement. The Jewish National Fund, historically tasked with large-scale afforestation efforts, is also engaged in both education and outreach, designing hiking trails and overseeing their use. The management of the core areas is the joint responsibility of the Megiddo Regional Council, the INPA, and the Yizraeleem Area Planning and Building Commission. The buffer area is under the joint management of the Jewish National Fund, the Grazing Authority of the Ministry of Agriculture, and the Yizraeleem Area Planning and Building Commission. A co-operative declaration outlines the responsibilities of each of the involved parties concerning transparency, conservation, education, and the promotion of the Reserve's goals.

The Megiddo Biosphere Reserve case study is an innovative example of a conservation function enforced by a civil agreement initiated and adopted by the local communities, with the support of additional statutory conservation instruments provided by local and state governments. The site also received support and guidance from UNESCO, and in 2011, it was accepted as an international Man and Biosphere Reserve (Ramat Menashe Biosphere Reserve Nomination, 2010).

6.3 The Shikma Park Plan

The Shikma Park's geographical features include seasonal floodplains, streams, sand dunes, *kurkar* ridges, springs and natural habitats, ancient orchards, recreational woodlands, plantings along the riverbeds, more recent groves and natural flora. The area is predominantly agricultural. Its landscapes feature field crops, groves, farms, and pastureland, creating a pastoral landscape dotted with rural communities. The park area has a rich cultural heritage recalling many periods of settlement: ancient *tels,* antiquities, historic buildings, remains of infrastructure from the British mandatory period, and Israeli national heritage sites and monuments.

6.3.1 Plan

The planning concept of Shikma Park is based on a continuum of natural and agricultural open spaces with varying levels of conservation protection. The park is treated as a single geographical and organizational unit that parallels the Shikma stream, including its tributaries, along with rural communities and recreational areas. The park crosses the jurisdictions of six regional councils, including Hof Ashkelon, Sha'ar HaNegev, Lachish, Merchavim, Bnei Shimon, and Shafir. This broad strip of open spaces between the foothills to the east and the Mediterranean Sea to the west drains into the Shikma stream basin. The plan's overarching goal is to protect the natural and cultural values of the agricultural landscape. The plan strives to expand the region's potential for sustainable tourism. It creates a platform for small local initiatives to serve the park with rural lodging options and other tourist services appropriate to the region's pastoral quality and sense of expansive space while protecting its local characteristics.

6.3.2 Plan Elements

One of the techniques used to define the park's boundaries is the use of a unified design language to mark the entrance points to the park. The same landscape treatment is used in the development of visitor services. A central path was designed to link the park's open spaces and enable access to points of interest and nearby local communities. The plan's open spaces have been ranked according to their sensitivity to development, using four levels:

Level 1—Edges of the park: The edges are primarily agricultural, with low to mid-level landscape values.

Level 2—Open park landscape: Open spaces primarily used for agriculture with mid to high-level landscape values.

Level 3—Buffer around the core conservation area: This is a continuous unit slated for conservation with high-level environmental values. The area contains *tels*, orchards, forests, heritage sites, lookout points, and places to observe seasonal wildflowers and agricultural landscapes.

Level 4—Park core: This unit is intended to preserve areas with very high-level landscape and ecological values, characterized by a continuous series of sensitive, open spaces requiring protection. This area contains nature reserves, forests, orchards, streams and bodies of water, and agriculture. Continuous ecological corridors have also been preserved throughout the length of the park.

The Shikma Park management administration is concerned with routine management, the visitor interface, and the park's promotion. This joint endeavor includes the regional councils with jurisdiction over land within the park's boundaries, public agencies, stakeholders, and representatives of local communities (Shikma Park Plan Guidelines, 2012).

6.3.3 Stakeholders

One of the plan's goals was to create a strong relationship between the local communities and open spaces. The communities were considered a critical and integral part of the park. The INPA, the Planning Administration, regional councils, environmental non-governmental organizations, the Israel Land Authority, and the general public were all involved in shaping the Shikma Park plan. A noteworthy element of the plan was the requirement for public participation. The INPA promoted public participation in the planning process so the residents could make their voices heard. Indeed, transparency was required as part of the public participation process; all data must be made available on the internet, including maps, reports, committee decisions, and additional materials to help the public make informed decisions.

References

Coetzer, K. L., Witkowski, E. T., & Erasmus, B. F. (2014). Reviewing biosphere reserves globally: Effective conservation action or bureaucratic label? *Biological Reviews, 89*(1), 82–104.

Incense Route UNESCO Nomination. (2005). Retrieved from https://whc.unesco.org/uploads/nominations/1107rev.pdfRamotMenasheBiosphereReserveNomination2010.

Megiddo Biosphere Reserve. Retrieved from https://en.unesco.org/biosphere/eu-na/megiddo.

Shikma Park Plan Guidelines (2012).

UNESCO. (1995). World Network of Biosphere Reserves: Statutory Framework UNESCO, Paris. Retrieved from: https://unesdoc.unesco.org/ark:/48223/pf0000373378.

UNESCO. (1996). Biosphere reserves: The Seville Strategy and the statutory framework of the world network. UNESCO, Paris. Retrieved from http://www.mab.cas.cn/ryswqjh/swqbhq/201411/W020141113678526165131.pdf.

UNESCO. (2021). Operational guidelines for the implementation of the world heritage convention, UNESCO World Heritage Centre, Paris. https://whc.unesco.org/en/guidelines/.

7 Conclusion

Over the last decade, the significance of cultural landscapes has become more widely understood and accepted in Israel. Population growth, accelerated development, growing concern about diminishing landscape resources, and fear of damage to cultural assets have all led to the recognition that natural landscape and cultural heritage assets are fundamentally related. This relationship defines the concept of "cultural landscape." Since 2010, four approaches can be identified in the Israeli discourse on cultural landscapes: the recognition of the historical and cultural dimensions of landscape; the ecological approach, which has underscored the need for landscape management; the economic approach; and the social approach.

These four approaches have had a significant impact on the Israeli perception of cultural landscapes, which represent not only the combined works of humans and nature but are also the places of peoples' livelihoods, identities, and beliefs. As such, our interest in cultural landscape stresses the social domain, allowing us to understand the influence of landscape on the cultural, political, and ideological realms.

The publication of the Hebrew edition of this volume in 2010 brought about a new understanding of cultural landscapes, along with the recognition of the importance of integrating this concept into Israeli landscape planning. The original volume succeeded in forging a broad consensus on the definition of cultural landscapes based on the international literature. One of the book's key goals was to provide information to national planning institutions to make cultural landscapes accessible and transparent to decision-makers and the relevant communities. Indeed, the issue has become part of the Israeli planning discourse and practice at all planning levels—national, district, and local—and has been adopted by many public agencies and local authorities. The book has been used by local authorities and public agencies, including the Israel Nature and Parks Authority, Antiquities Authority, Council for the Conservation of Heritage Sites, academic institutions, local authorities, and other government agencies. Its publication contributed to a new awareness of cultural landscapes in the education systems and university curricula. It strengthened the recognition of world definitions of the cultural landscape and promoted comparative research between Israel and other countries worldwide. It also encouraged stakeholder participation in the decision-making process and the design of cultural landscapes to reduce the tension between development and conservation.

Yet, this is just a beginning. Additional research is needed to further develop the concept of cultural landscapes and expand their role in protecting open spaces in Israel, providing livelihood for the local communities and quality of life for all. Unlike other countries where development is generally more gradual, the Israeli landscape is being transformed at an accelerating pace. There is an urgent need to strengthen the regulatory system to protect cultural landscapes by facilitating their designation and incorporating them into existing planning mechanisms. Public discussions regarding their value and contribution to well-being, livelihoods, and the environment must also be carried out more vigorously and inclusively. For example, Israel is known for its development trends threatening the continued existence of open space, natural resources, and cultural heritage sites in rural areas. These trends are evident in many countries worldwide, yet they are especially conspicuous and threatening in Israel, a small and densely populated country with limited land resources. Moreover, the present urban-biased development trends pose a threat to the continued existence of Israeli rural cultural landscapes, including the settlement models of the kibbutz and moshav, which are unique and, therefore, highly valued cultural heritage assets. A "missing link" is required to create an integrated planning approach to conserving rural cultural landscapes, settlements, and agricultural lands (Amit-Cohen, 2018; Amit-Cohen & Sofer, 2017).

Since the book's initial publication, dramatic shifts have occurred in how cultural landscapes have been conceptualized and implemented globally. The climate crisis is perhaps

A. Sar Shalom et al., *Cultural Landscapes of Israel*,
https://doi.org/10.1007/978-3-031-33685-0_7

the paramount challenge to planning theory and practice. The second challenge is social and political, arising from the demand for more pluralistic and often activist perspectives. The following discussion highlights the key issues associated with these two themes that Israel needs to address in the coming years.

7.1 Nature and Culture: The Climate Crisis, Agriculture, and Food Security

While cultural landscapes purport to link the cultural with the natural, these realms remain stubbornly divergent. The nature-culture divide is perhaps one of the most profound challenges to cultural landscape conservation policy, and the climate crisis has made this problem even more acute. How do we move toward a more holistic vision in which culture and nature are "constituted relationally"? (Larsen & Wijesuriya, 2017, p. 145). This requires "new institutional practices, and a new language to address interconnected natural and cultural values" (Larsen & Wijesuriya, 2017, p.149). Spiritual values, cultural conservation practices, traditional ecological management knowledge, and stewardship practices are all interlinked and help maintain a cultural landscape's integrity. The second question relates to scale: research and planning must be undertaken at a territorial scale to link cultural heritage with natural resources (Dastgerdi et al., 2019) and bridge the gap between the study of heritage and cultural ecosystem services. The case studies of the Megiddo Biosphere Reserve and Shikma Park (see Chap. 6) are harbingers of this trend.

Such a vision offers a synergistic approach to planning in which local communities protect and manage cultural landscapes as an integral part of territorial and spatial planning (Aktürk & Dastgerdi, 2021). This perspective brings a more proactive agenda to cultural landscape planning, shifting away from the narrowly reactive approach that characterized early thinking on conserving World Heritage sites. It advocates for "adopting a territorial planning approach for the protection of cultural heritage rather than a traditional conservation approach, through a more comprehensive and accurate understanding of the relationship between cultural heritages the natural resources" (Dastgerdi et al., 2019, p. 7).

Linking natural and cultural values requires integrative tools and mechanisms at every stage of the process: from assessing values to developing management programs. It requires institutional frameworks that encourage cooperation across distinct domains of expertise (Larsen & Wijesuriya, 2017). In Israel, cultural landscapes have had a significant impact on the planning frameworks, especially by shaping the idea of the "landscape ensemble" that has been a key feature of the National Outline Plan 35 since 2016 and the 2020 integrative National Outline Plan 1 (see Chap. 4). Collaborative planning and shared expertise will be critical in the future to advance a holistic agenda bridging natural and cultural values.

Climate change adaptation has only recently been addressed within the cultural heritage field.[1] It must be addressed as an integral component of cultural landscape protection and management, including at archaeological sites (Cassar, 2016; Sesana et al., 2018). To date, there are few documented cases of adaptation of cultural heritage to climate change, though there is increasing awareness of the need to adopt more proactive, as opposed to reactive measures (Sesana et al., 2018). Emergency preparedness and disaster risk reduction also need to be addressed, along with methods of monitoring landscape systems (Dastgerdi et al., 2019; Sánchez et al., 2020). In Israel, mitigating the effects of flash storms and flooding is a high priority, together with preventing coastal cliff erosion.

Agricultural landscapes play an essential role in this context in addition to their role in promoting food security. This is especially true in Israel, where cultivation is declining dramatically every year due to rising development pressures. Moreover, changing agricultural technologies are affecting the physical environment. In her background paper for a workshop on Globally Important Agricultural Heritage Systems (GIAHS), Mechtild Rossler has noted that "agricultural and rural landscapes may be one of the main themes for landscape identification and future regional and national reviews may be also carried out keeping in mind that these landscapes contribute to the food production of the world" (Rossler, n.d., p. 8). Sustainable agriculture is related to stakeholders' cultural practices and livelihoods, as well as promoting agrodiversity, which the Convention on Biological Diversity defines as preserving genetic resources for food and agriculture and biodiversity that supports ecosystem services of agriculture (United Nations Environmental Programme, 1992; Zimmerer, 2014). The communities adjacent to the cultural landscapes become guardians and may assume a management role with new economic potential.

7.2 Pluralistic Planning and Contested Landscapes

It has been written that cultural heritage policies and practices are inherently political (Silverman, 2011, p. 24). What is deemed worthy of protection, and what is erased? Which narratives are valued, and which are silenced? And who decides? These questions take on added intensity in a society

[1] According to Sesana et al. (2018), the impacts of climate change on cultural heritage were first mentioned in 2014 in a report by the Intergovernmental Panel on Climate Change (IPCC).

defined by conflict. Israel's landscapes contain the rituals, meanings, values, and memories of its two main cultures: Jewish and Arab. Cultural heritage in Israel often directly serves national goals, cementing certain narratives and denying others. The creation of a more pluralistic landscape is an ongoing challenge.

This can be facilitated by promoting broader civic engagement with diverse communities and stakeholder groups to allow multiple narratives to emerge. Contested landscapes can be addressed in two ways. The first is to create a common environment for more than one cultural or interest group, within which each group has its separate cultural-symbolic landscape. Each group, thus, possesses the territory of its symbolic landscapes. The second approach is to create an environment containing a joint cultural landscape for multiple groups, but to which each group ascribes different significance and purpose. When a landscape is contested, or when there is a struggle between two groups, with each group claiming ownership of the landscape, a struggle also ensues between the different icons of the groups. If one group takes over the contested area, the result may be a deletion of the symbols and icons of the group evicted from the disputed territory.

Much work remains to be done. In Israel, there is an urgent need to continue making the planning process more holistic and flexible, and to develop a more sophisticated toolkit for managing landscapes, reconciling nature and culture, and opening up the conversation to more diverse voices. Like a written text, the landscape has many readers and, therefore, can be interpreted on many levels. The different interpretations can lead to disputes and arguments but may also advance awareness. It is our hope that the study of the cultural landscapes in Israel will deepen cultural understanding and help lead to coexistence.

References

Aktürk, G. & Dastgerdi, A. S. (2021). Cultural landscapes under the threat of climate change: A systematic study of barriers to resilience. *Sustainability, 13*, 9974.

Amit-Cohen, I. (2018). Conservation rural space–The case of agricultural cooperative settlements and open space in Israel. *Heritage Architecture Studies*, 165.

Amit-Cohen, I. A., & Sofer, M. (2017). Integrated rural heritage landscapes: The case of agricultural cooperative settlements and open space in Israel. *Journal of Rural Studies, 54*, 98–110.

Cassar, J. (2016). Climate change and archaeological sites: Adaptation strategies. In R.-A. Lefèvre & C. Sabbioni (eds.), *Cultural heritage from pollution to climate change*.

Dastgerdi, A. S., Sargolini, M., & Pierantoni, I. (2019). Climate change challenges to existing cultural heritage policy. *Sustainability, 11* (19), 5227.

Larsen, P. B., & Wijesuriya, G. (2017). Nature–Culture interlinkages in world heritage: Bridging the gap. In *The George Wright Forum* (Vol. 34, pp. 142–153). JSTOR.

Rössler, M. (undated) UNESCO World Heritage Centre background document on UNESCO World Heritage Cultural Landscapes Prepared for the FAO Workshop and Steering Committee Meeting of the GIAHS project: Globally Important Ingenious Agricultural Heritage Systems.

Sánchez, M. L., Cabrera, A. T., & Del Pulgar, M. L. G. (2020). Guidelines from the heritage field for the integration of landscape and heritage planning: A systematic literature review. *Landscape and Urban Planning, 204*, 103931.

Sesana, E., Gagnon, A. S., Bertolin, C., & Hughes, J. (2018). Adapting cultural heritage to climate change risks: Perspectives of cultural heritage experts in Europe. *Geosciences, 8*(8), 305.

Silverman, H. (2011). Contested cultural heritage: A selective historiography. In *Contested cultural heritage* (pp. 1–49). Springer.

United Nations Environment Programme (1992). Convention on biological diversity. https://wedocs.unep.org/20.500.11822/8340.

Zimmerer, K. S. (2014). Conserving agrobiodiversity amid global change, migration, and nontraditional livelihood networks: The dynamic uses of cultural landscape knowledge. *Ecology and Society, 19*(2).

Correction to: Case Studies

Correction to:
Chapter 5 in: A. Sar Shalom et al., *Cultural Landscapes of Israel*,
https://doi.org/10.1007/978-3-031-33685-0_5

In the original version of the book, the following belated corrections have been incorporated in chapter "Case Studies":

1. In page 46, the name "Rafi Lewis" has been changed to "Rafael Y. Lewis" in the footnote.
2. In page 78, the following footnotes has been included: This section was written by Nurit Lissovsky.

The book and the chapter have been updated with the changes.

The updated version of this chapter can be found at
https://doi.org/10.1007/978-3-031-33685-0_5

A. Sar Shalom et al., *Cultural Landscapes of Israel*,
https://doi.org/10.1007/978-3-031-33685-0_8

Appendix A

Criteria for Outstanding Universal Value

(Operational Guidelines for the Implementation of the World Heritage Convention, 2021, Paragraph 77).

The Committee considers a property as having Outstanding Universal Value (see paragraphs 49–53) if the property meets one or more of the following criteria. Nominated properties shall therefore:

(i) represent a masterpiece of human creative genius;
(ii) exhibit an important interchange of human values, over a span of time or within a cultural area of the world, on developments in architecture or technology, monumental arts, town-planning, or landscape design;
(iii) bear a unique or at least exceptional testimony to a cultural tradition or to a civilization that is living or which has disappeared;
(iv) be an outstanding example of a type of building, architectural or technological ensemble, or landscape that illustrates (a) significant stage(s) in human history;
(v) be an outstanding example of a traditional human settlement, land-use, or sea-use, which is representative of a culture (or cultures), or human interaction with the environment especially when it has become vulnerable under the impact of irreversible change;
(vi) be directly or tangibly associated with events or living traditions, with ideas, or with beliefs, with artistic and literary works of outstanding universal significance (The Committee considers that this criterion should preferably be used in conjunction with other criteria);
(vii) contain superlative natural phenomena or areas of exceptional natural beauty and aesthetic importance;
(viii) be outstanding examples representing major stages of earth's history, including the record of life, significant on-going geological processes in the development of landforms, or significant geomorphic or physiographic features;
(ix) be outstanding examples representing significant on-going ecological and biological processes in the evolution and development of terrestrial, freshwater, coastal, and marine ecosystems and communities of plants and animals;
(x) contain the most important and significant natural habitats for in-situ conservation of biological diversity, including those containing threatened species of Outstanding Universal Value from the point of view of science or conservation.

These criteria were formerly presented as two separate sets of criteria—criteria (i)–(vi) for cultural heritage and (i)–(iv) for natural heritage. The 6th extraordinary session of the World Heritage Committee decided to merge the ten criteria (Decision 6 EXT.COM 5.1).

Thematic studies are different than the comparative analysis to be prepared by States Parties when nominating properties for inscription in the World Heritage List (see paragraph 132). Decision 43 COM 11A Decision 43 COM 11A Operational Guidelines for the Implementation of the World Heritage Convention 30.

A. Sar Shalom et al., *Cultural Landscapes of Israel*,
https://doi.org/10.1007/978-3-031-33685-0

Appendix B

Israel World Heritage Listing

World Heritage properties inscribed

- Masada
- Old City of Acre
- White City of Tel-Aviv—The Modern Movement
- Biblical Tels—Megiddo, Hazor, Beer Sheba
- Incense Route—Desert Cities in the Negev
- Bahá'i Holy Places in Haifa and the Western Galilee
- Sites of Human Evolution at Mount Carmel: The Nahal Me'arot/Wadi el-Mughara Caves
- Caves of Maresha and Bet-Guvrin in the Judean Lowlands as a Microcosm of the Land of the Caves
- Necropolis of Bet She'arim: A Landmark of Jewish Renewal.

World Heritage properties noted on the Tentative List
Last Revision—05/02/2015

- Arbel (Arbel, Nebe Shueb, Horns of Hattin)
- Bet She'an
- Caesarea
- Degania & Nahalal
- Early Synagogues in the Galilee
- Ein Karem, a Village and its Cultural Landscape
- Horvat Minnim
- Jerusalem*
- Liftah (Mey Naftoah)—Traditional Mountain Village
- Makhteshim Country
- Mount Karkom
- Sea of Galilee and its Ancient Sites
- The Crusader Fortresses
- The Galilee Journeys of Jesus and the Apostles
- The Great Rift Valley—Migratory Routes—The Hula
- *Timna
- Triple-arch Gate at Dan and Sources of the Jordan
- White Mosque in Ramle.

A. Sar Shalom et al., *Cultural Landscapes of Israel*,
https://doi.org/10.1007/978-3-031-33685-0

Appendix C

The Cultural Landscape Geographic Data Table

1. Cultural Landscapes with a Universal Level of Significance

1.1 Inscribed on the World Heritage List as Cultural Landscape

Name of the Cultural Landscape	Area (in hectares)	Cultural Anchors	Buffer Zone (in hectares)
The Incense Route	6314.2	Moa Caravanserai	56277.3
		The Nekarot Stronghold	
		The Saharonim Caravanserai	
		Mahmal Stronghold	
		Avdat	
Ancient Agriculture in the Negev	3977.8	Moshe Cistern and Aharon Cistern Nitzana Milka Cistern	
	6861.0	Avdat	
	10490.6	Shivta	

1.2 Inscribed on the World Heritage List

Name of the Cultural Landscape	Area (in hectares)	Cultural Anchors	Buffer Zone (in hectares)
The Bahá'í Gardens	19.9	Al Bahaja (House of Bahá'u'lláh), Acre	33.5
	30.5	Shrine of the Báb (The Golden Dome), Haifa	25.6
Megiddo and Armageddon—Ramat Menashe	23	Tel Megiddo	3890
Land of the Caves	5136.6	Khirbet Atari	12593.6
		Khirbet Burgin	
		Khirbet Refa	
		Khirbet Resek	
		Beit Guvrin	

A. Sar Shalom et al., *Cultural Landscapes of Israel*,
https://doi.org/10.1007/978-3-031-33685-0

1.3 Recorded on the World Heritage Tentative List

Name of the Cultural Landscape	Area (in hectares)	Cultural Anchors	Buffer Zone (in hectares)
*Christian Sites in the Galilee (Jesus and the Apostles)	31,767	Kefar Kana	
		Roman Road Remains at the Golani Junction	
		Arbel Cliffs	
		Magdala	
		Tabgha	
		Capernaum	
		Korazim	
		Mount of Beatitudes	
		Bethsaida	
		Kursi	
		Tel Hadar	
Mount Tabor—the Transfiguration	362.2	Franciscan Church	213.5
		Orthodox Church	
The Horns of Hattin	286.4	Maskana Pool	
		The Watershed	
		The Plain of Hattin	
		Coaxial Field System	
		Horns of Hattin	
Mount Karkom (Har Karkom)	709.2	Concentration of Petroglyphs on Mount Karkom	7016.8
		Karkom Well/Be'er Karkom	
Timna	6820.4	Temple of Hathor	7553.2
		Timna Ruins	
		Ancient Mines	
		Ancient Petroglyphs	
The Hula	2155	The Hula Reserve	5718
		The Hula Lake	
Nahalal	438.8		652.9
Crusader Structures in the Western Galilee	12.7	Mi'ilya Fortress	3899.4
	4.9	Montfort Castle	
	107.3	Yehi'am Fortress	
	3.9	Khirbet Manot	
Port Towns along the Carmel Coast	48.8	The Dor Port	
	165.2	The Caesarea Port	5367.6
	51.2	Atlit Yam (Atlit by the Sea)	
Terraced Agriculture in the Judean Mountains	948.1	Irrigation Farming in Refaim Stream	
		Ein Lavan	
		Ein Al-Hania	
		Ein Al-Balad	
		Ein Carmon	
	210.7	Ein Bikura (Sataf)	
	303.8	Ein Heret	
	336.9	Ein Tzuba	
Traditional Mountain Village		Liftah	36.6

2. Cultural Landscapes with a National Level of Significance

2.1 Designed Landscapes

Name of the Cultural Landscape	Area (in hectares)	Cultural Anchors	Buffer Zone (in hectares)
Sde Refaim—	77.5	The Dolmens	
	387.2	Khirbet Berekh	
	12.6	Gilgal Refaim, Rujm el-Hirh	781.8
	397.9	Ramat Gamla	
Gan HaShlosha/The Sachne	13		16
Avenues of Washingtonia Palms	3.9	The Aaronsohn Experimental Station structure	41.7
	8.5	Hulda Forest, Herzl House	74.7
	2.3	Sde Eliyahu	15.4
HaNadiv Heights (Ramat HaNadiv)	9.7	The Tomb of the Baron de Rothschild	442.2
The Ben-Gurion Tomb	3.9	The Ben-Gurion tomb	
Liman Irrigation Systems in the Negev	0.6	Be'er Chayal	
	2.2	Haro'a Picnic Area	
	0.3	Ramat Boker	

2.2 Organically Evolved Landscapes: Relict

Name of the Cultural Landscape	Area (in hectares)	Cultural Anchors	Buffer Zone (in hectares)
Circassian Villages in the Golan	26.9	Surman Ruins	206.6
Agricultural Walls in the Central Golan	61.9		849
Ancient Rural Settlement in the Golan	19.2	Khirbet Faraj	225.3
Grove Houses (Be'arot)	0.4	Khan Zarkaniyya	1.9
	0.7	Burj Binyamina	10.6
	0.4	Bitan Aharon	5.6
Grove Houses Estates	9.5	The Gan Raveh Estate	1134.4
	2.2	Gurland Well (Be'er Gurland)	
	2.3	Mordechai Farm	
British Ammunition Depot	168.4		
The Workers' Camp at Sodom	10.2	Workers' living quarters	
		The dining hall	
		The Colonel's cave	
Wilhelma (Bnei Atarot)	60.6		146.1
Monasteries in the Judean Lowlands (Shfela)	36.5	Dir Rafat Monastery	1600.3
	4.5	Emmaus Monastery	
	65.7	Latrun Monastery	
A Row of *Tels* in the Judean Lowlands (Shfela)	46.2	Tel Tzora	23105.1
	31.5	Tel Beit Shemesh	
	129.6	Tel Azekah	
	54.5	Tel Goded	
	69.2	Tel Maresha	
	30.1	Tel Lachish	
Rows of Sycamore Trees	633.9	Rows of Sycamore Trees	
The Scorpion's Ascent	608.8	The Engineering Corps Memorial	
		Tzafir Stronghold	
		Tzafir	
		Rujm Tzafir	

2.3 Organically Evolved Landscapes: Continuing

Name of the Cultural Landscape	Area (in hectares)	Cultural Anchors	Buffer Zone (in hectares)
Olive Groves in the Galilee	1129.5	Maghar	
	9.97	Rama—Hanania Farm	
Beit Netofa Valley	3003.9		3540.8
Irrigated Agriculture in the Harod Valley (Emek Harod)	4157.2		
HaNadiv Valley	3225.4		
The Groves of the Sharon Plain	2309.1	The Zeta Swamp	
		Tel Zeror	
Bat Shlomo	12.6		36.8
Mikve Israel	94.5		529.6
Deciduous Groves in the Ya'afuri Valley	983.4	Birkat Ram (Lake Ram)	
		Nabi Ya'afuri (Tomb of the Ya'afuri Prophet)	

2.4 Associative Landscapes

Name of the Cultural Landscape	Area (in hectares)	Cultural Anchors	Buffer Zone (in hectares)
Tombs of the Tzadikim in the Galilee	5295.5	Shma'ya and Avtalyon	35.7
		Honi Hama'agel and his wife	
		Rabbi Elazar HaModai and Jacob Ish Navoriah	
		Rabbi Bana'ah the Amora	
		Rabbi Hillel the Amora, Judah Ben Tima	
		Rabbi Yehonatan Ben Uziel	
		Rabbi Yossi the Galilean, Eliyahu the Cohen Gaon	
		Rabbi Yosef Ben Zimra	
		Rabbi Crosfodai the Amora	
The Tomb of Rabbi Shimon Bar Yohai	0.4	The Tomb of Rabbi Shimon Bar Yohai	
		The ancient synagogue in Meron	
The Top of Mount Carmel (Rosh HaCarmel)	62.5	The cave of the prophet Elijah	
		Stella Maris Monastery	
		The Lighthouse	
"The Guard of the Fields": Alexander Zaïd ("Al Mishmar HaSadot")	392.3	The statue of Alexander Zaïd	
		The cemetery of the Shomer organization	
King Saul and the Philistines at the Gilboa	3597.2	Moreh Hill	
		EinDor	
		Tel Jezreel	
		Mount Saul (Har Sha'ul)	
		Harod Spring	
Gate of the Valley (Sha'ar HaGai)	297.8	Khirbet Harsis	451.9
		The Sha'ar HaGai Khan	
Elijah and the Prophets of Baal	1457	Keren HaCarmel	
		Metzudat Habamah (Khirbet Kliaa)	
David and Goliath	30.1	Tel Shoka	5831
	129.6	Tel Azekah	
	114.4	Tel Tzafit	

Glossary

Be'er well (Hebrew). Many wells were named, i.e., Be'er Miriam translates as Miriam's Well.

Be'ara (plural: *be'arot*) well house (Hebrew and Arabic) or building compound constructed near a well to support orange groves in the nineteenth century in the coastal areas of Palestine.

Beit house (Hebrew and Arabic). Many locations have *beit* preceding the name of the place, including Beit Guvrin, Beit Shean, and Beit Netofa.

Bikat "valley of" (Hebrew) i.e., Bikat HaNadiv is translated as "valley of the generous," alluding to the benefactor Baron de Rothschild.

Burj tower (Arabic). Several places in Israel have *burj* preceding the name of the place, e.g., Burj Binyamina.

Ein spring (Arabic and Hebrew).

Gan garden or park (Hebrew), e.g., Gan Hashlosha (Park of the Three)

Har mountain (Hebrew). e.g., Har Tavor, *Har Karkom* is Mount Karkom.

Hectare A metric system unit used in the measurement of land. Hectares were created as a metric replacement for the imperial acre. A hectare contains about 2.47 acres.

Khan (Turkic). In the Middle East the term *khan* has two meanings and can refer to a roadside inn or an inner-town inn. Khan Sarahonim and Khan Sha'ar Hagai are two examples of roadside inns.

Khirbet ruin (Arabic). The term is found commonly found in the names of ancient and other long abandoned sites throughout Israel and the Middle East.

Kibbutz (plural: *kibbutzim*). A communal settlement in which all production means, land rights, homes, and community institutions are collectively owned. The first kibbutz was Degania, established in 1909. Though traditionally based on agriculture, most kibbutzim contain industry. Today most kibbutzim have undergone varying degrees of privatization.

Nahal stream or river (Hebrew).

Tel (Hebrew and Arabic). A type of archaeological mound created by human occupation and abandonment of a geographical site over many centuries. A classic *tel* looks like a low, truncated cone with a flat top and sloping sides.

Makhtesh crater (Hebrew)—A geological landform found in the Negev desert of Israel and the Sinai Peninsula of Egypt. A *makhtesh* has steep walls of resistant rock surrounding a deep closed valley usually drained by a single wadi. The valleys have limited vegetation and soil, containing a variety of different colored rocks and diverse fauna and flora.

Mikve (Hebrew) Ritual baths used in the Jewish tradition for ritual immersion and purification. A *mikve* is a pool of still water that comes from a natural source.

Moshav, Moshava (plural: *moshavot*). A cooperative agricultural community of independent farmers pioneered by the Labor Zionists between 1904 and 1914.

A. Sar Shalom et al., *Cultural Landscapes of Israel*,
https://doi.org/10.1007/978-3-031-33685-0

Ramat heights (Hebrew). A geographical reference to the site's topography and indicates a higher elevation relative to its surroundings.

Regional Council Regional Councils are a type of local government in Israel that represent small communities.

Rujm mound, cairn. or tumulus (Arabic). Rujm el-Hiri is an example of a site with dolmens.

Sha'ar gate (Hebrew). The word also has geographic associations and can allude to the entry to a region, e.g. Sha'ar HaGai, which is translated as the "gate of the valley".

MIX
Papier aus verantwortungsvollen Quellen
Paper from responsible sources
FSC® C105338

If you have any concerns about our products,
you can contact us on
ProductSafety@springernature.com

In case Publisher is established outside the EU,
the EU authorized representative is:
Springer Nature Customer Service Center GmbH
Europaplatz 3, 69115 Heidelberg, Germany

Printed by Libri Plureos GmbH
in Hamburg, Germany